my **revision** notes

dexcel AS
GEOGRAPHY

Michael Raw

Hodder Education, an Hachette UK company, 338 Euston Road, London NW1 3BH

Orders

Bookpoint Ltd, 130 Milton Park, Abingdon, Oxfordshire OX14 4SB

tel: 01235 827827

fax: 01235 400401

e-mail: education@bookpoint.co.uk

Lines are open 9.00 a.m.–5.00 p.m., Monday to Saturday, with a 24-hour message answering service. You can also order through the Hodder Education website: www.hoddereducation.co.uk

First printed 2013

Impression number 5 4 3

Year 2017 2016 2015

Cover photo reproduced by permission of freshidea/Fotolia

Typeset by Datapage (India) Pvt. Ltd.

Printed in Spain

Hachette UK's policy is to use papers that are natural, renewable and recyclable products and made from wood grown in sustainable forests. The logging and manufacturing processes are expected to conform to the environmental regulations of the country of origin.

P2183

Get the most from this book

Everyone has to decide his or her own revision strategy, but it is essential to review your work, learn it and test your understanding. These Revision Notes will help you to do that in a planned way, topic by topic. Use this book as the cornerstone of your revision and don't hesitate to write in it — personalise your notes and check your progress by ticking off each section as you revise.

☑ Tick to track your progress

Use the revision planner on page 4 to plan your revision, topic by topic. Tick each box when you have:

● revised and understood a topic
● tested yourself
● practised the exam questions and gone online to check your answers and complete the quick quizzes

You can also keep track of your revision by ticking off each topic heading in the book. You may find it helpful to add your own notes as you work through each topic.

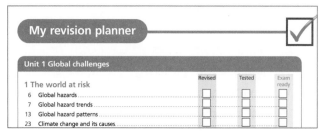

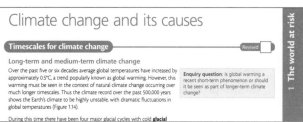

Features to help you succeed

Examiner's tips and summaries

Expert tips are given throughout the book to help you polish your exam technique in order to maximise your chances in the exam.

The summaries provide a quick-check bullet list for each topic.

Typical mistakes

The author identifies the typical mistakes candidates make and explains how you can avoid them.

Now test yourself

These short, knowledge-based questions provide the first step in testing your learning. Answers are at the back of the book.

Definitions and key words

Clear, concise definitions of essential key terms are provided on the page where they appear. Key words from the specification are highlighted in bold for you throughout the book.

Exam practice

Practice exam questions are provided for each topic. Use them to consolidate your revision and practise your exam skills.

Online

Go online to check your answers to the exam questions and try out the extra quick quizzes at **www.hodderplus.co.uk/myrevisionnotes**

My revision planner

Unit 1 Global challenges

Unit 2 Geographical investigations

Countdown to my exams

6–8 weeks to go

- Start by looking at the specification — make sure you know exactly what material you need to revise and the style of the examination. Use the revision planner on page 4 to familiarise yourself with the topics.
- Organise your notes, making sure you have covered everything on the specification. The revision planner will help you to group your notes into topics.
- Work out a realistic revision plan that will allow you time for relaxation. Set aside days and times for all the subjects that you need to study, and stick to your timetable.
- Set yourself sensible targets. Break your revision down into focused sessions of around 40 minutes, divided by breaks. These Revision Notes organise the basic facts into short, memorable sections to make revising easier.

Revised ☐

4–6 weeks to go

- Read through the relevant sections of this book and refer to the examiner's tips, examiner's summaries, typical mistakes and key terms. Tick off the topics as you feel confident about them. Highlight those topics you find difficult and look at them again in detail.
- Test your understanding of each topic by working through the 'Now test yourself' questions in the book. Look up the answers at the back of the book.
- Make a note of any problem areas as you revise, and ask your teacher to go over these in class.
- Look at past papers. They are one of the best ways to revise and practise your exam skills. Write or prepare planned answers to the exam practice questions provided in this book. Check your answers online and try out the extra quick quizzes at **www.hodderplus.co.uk/myrevisionnotes**
- Try different revision methods. For example, you can make notes using mind maps, spider diagrams or flash cards.
- Track your progress using the revision planner and give yourself a reward when you have achieved your target.

Revised ☐

One week to go

- Try to fit in at least one more timed practice of an entire past paper and seek feedback from your teacher, comparing your work closely with the mark scheme.
- Check the revision planner to make sure you haven't missed out any topics. Brush up on any areas of difficulty by talking them over with a friend or getting help from your teacher.
- Attend any revision classes put on by your teacher. Remember, he or she is an expert at preparing people for examinations.

Revised ☐

The day before the examination

- Flick through these Revision Notes for useful reminders, for example the examiner's tips, examiner's summaries, typical mistakes and key terms.
- Check the time and place of your examination.
- Make sure you have everything you need — extra pens and pencils, tissues, a watch, bottled water, sweets.
- Allow some time to relax and have an early night to ensure you are fresh and alert for the examination.

Revised ☐

My exams

AS Geography Unit 1

Date: ....

Time: ....

Location: ....

AS Geography Unit 2

Date: ....

Time: ....

Location: ....

1 The world at risk

Global hazards

Natural hazards and natural disasters

Revised

Natural hazards are extreme **hydro-meteorological** and **geophysical** events, such as cyclones, floods, earthquakes and volcanic eruptions, which impact adversely on people, economy and society (Figure 1.1). They invariably result in death or injury, the destruction of property and infrastructure, and significant economic and social disruption.

Natural events only *become* hazards when they interact with people. For instance, a volcanic eruption on an uninhabited island is, by definition, non-hazardous. When natural hazards cause severe loss of life, injury and economic damage, they are known as **natural disasters**.

Enquiry question: What are the main types of physical risk facing the world and how big are they?

Now test yourself
Tested

1   When does a natural hazard become a natural disaster?
2   When does an earthquake or a cyclone become a hazard?

Answers on p. 121

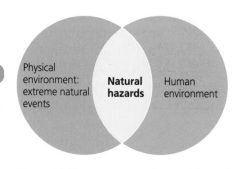

Figure 1.1 The occurrence of natural hazards

Measuring disaster risk

Revised

Disaster risk describes the probability of a disaster's occurrence and the harmful consequences that are likely to result. This is summarised by the **disaster equation**:

$$R = \frac{H \times V}{C}$$

where:

R = risk of disaster

H = size/scale/probability of a hazard

V = vulnerability (e.g. levels of development, poverty, assets) and the number of people living in areas at risk

C = capacity to cope, i.e. emergency response, recovery capability, resistance to shock

Risk is therefore directly proportional to the:

● size and severity of a hazardous event

● number of people living in the affected area

● **vulnerability** of the people, economy and society in the affected area

Risk is inversely proportional to society's **preparedness** and ability to cope with natural hazards.

> **Examiner's tip**
>
> Any assessment of the risks posed by natural hazards must identify the magnitude and nature of the hazard, the number of people at risk, the vulnerability of the society, and the ability of society to respond and mitigate hazard impacts.

> **Typical mistake**
>
> Extreme natural events are by themselves not hazards. Natural events such as floods and volcanic eruptions only become hazardous when they adversely impact people and human activity.

Now test yourself

3 What factors contribute to the vulnerability of populations to natural hazards?

4 What is meant by the term 'disaster risk'?

Answers on p. 121

Global warming: the world's number one problem?

Arguably, global warming is the greatest problem currently facing humankind. This is because:

● global warming, by definition, is a problem on a global scale, affecting every part of the planet

● global warming has far-reaching impacts on environmental and economic systems, e.g. habitat change, mass plant and animal extinctions, food production, water supply

● rising sea level directly threatens many of the world's largest cities, millions of people living at or near sea level, and entire countries (e.g. Bangladesh, the Maldives)

● drought and the dislocation of human population caused by global warming could trigger unprecedented international migrations and political conflicts (e.g. competition for food and water resources)

● higher temperatures could spread infectious diseases from the tropics and sub-tropics to temperate latitudes (e.g. malaria)

● global warming could lead to shortages of water and food, as rainfall decreases in some places, glaciers melt and crop yields decline

● global warming is likely to increase the frequency and intensity of tropical cyclones, mid-latitude storms and floods

Global hazard trends

Hazards increasing in magnitude and frequency

Some natural hazards, such as river and coastal floods, tropical cyclones and droughts, appear to be increasing in frequency and magnitude. The underlying causes are a combination of physical and human factors.

Enquiry question: How and why are natural hazards now becoming seen as an increasing global threat?

Floods

There is evidence that the number of flood disasters has increased significantly since 1980 (Figure 1.2). In recent years there have been major river floods in Thailand (2011), USA (Mississippi 2011), south and central China (2011) and Pakistan (2010). Extreme rainfall, linked to tropical cyclones, **mid-latitude depressions** and thunderstorms, was responsible. This trend is consistent with:

● global warming, which increases evaporation and speeds up the atmosphere's general circulation

● population growth and development on floodplains, which has raised levels of vulnerability to flood hazards

● widespread deforestation due to population growth and pressure on resources, which has increased runoff and heightened flood risks

Now test yourself

5 Give two reasons why flood disasters appear to be increasing.

Answer on p. 121

Coastal flooding in the tropics and sub-tropics is often caused by storm surges driven by powerful cyclones. In 2005, the storm surge generated by Hurricane Katrina killed more than 1,100 people in New Orleans and along the Gulf of

Mexico. A similar storm surge accompanied Cyclone Nargis in May 2008; it swept through the Irrawaddy Delta in Burma, killing an estimated 140,000 people.

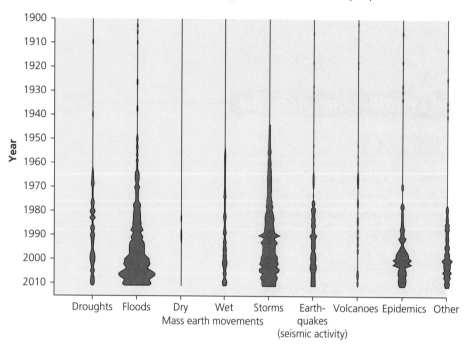

Figure 1.2 Number of natural disasters reported 1900–2010 (Source: EM-DAT, www.emdat.be)

Cyclones

The 2004 and 2005 hurricane seasons in the North Atlantic–Caribbean area produced a record-breaking number of storms, including five category 5 hurricanes (i.e. the most powerful) (Table 1.1). Hurricane Katrina was one of these storms. It caused more than US$100 billion in damage and more than 1,100 deaths. In the past 40 years, global warming has raised ocean surface temperatures by 0.6°C and has increased evaporation. The result is an increase in the power and frequency of tropical cyclones.

Even so, other factors influence the number of storms in any one year. Most important are the ocean surface **temperature anomalies** in the Pacific, known as **El Niño** and **La Niña** or the **El Niño Southern Oscillation** (ENSO).

El Niño suppresses tropical cyclone activity in the North Atlantic–Caribbean area. In contrast, La Niña (e.g. 2005) creates favourable conditions for hurricane development.

Climate models predict an increase in the frequency of mid-latitude cyclones (or **depressions**) in the UK in the next 30–40 years. This is due to global warming (and the melting of Arctic sea ice) and a southward shift of the main storm track of Atlantic depressions.

Droughts

Droughts are prolonged periods of abnormally low rainfall. They develop slowly and ultimately result in water shortages. The impact of drought is felt first by activities most dependent on rainfall such as arable farming, ranching, river transport and power supplies. Severe droughts may threaten public water supplies and exceptionally low river flow may cause long-lasting damage to aquatic ecosystems. Large parts of the UK suffered droughts in 1976, 1995 and 2012. In the Sahel region of Africa, prolonged droughts over the past 40 years have accelerated desertification and land degradation, and caused famine, migration and high levels of mortality. More than 1 million people have died in drought disasters in Africa since 1970.

In future, droughts are likely to become more frequent as climate change increases rainfall variability (Figure 1.3). The impact will be most strongly felt in

Table 1.1 Number of hurricanes in North Atlantic and Caribbean 2000–2011

Year	Number of hurricanes	Year	Number of hurricanes
2000	8	2006	5
2001	9	2007	6
2002	4	2008	8
2003	7	2009	3
2004	9	2010	12
2005	15	2011	7

Examiner's tip

The relationship between ocean temperatures and the world's climate is complex and not fully understood. When discussing trends in the frequency and magnitude of tropical cyclones it should be emphasised that in the short-term (i.e. year-to-year) there is considerable uncertainty and variability.

Typical mistake

Famines occur when there are food shortages and often cause sharp rises in mortality. However, it is wrong to assume that famines always imply an absolute shortage of food. There may be a breakdown in distribution or marketing systems, or food prices may be unaffordable for the poorest people.

Exam practice answers and quick quizzes at **www.hodderplus.co.uk/myrevisionnotes**

semi-arid and seasonally arid regions such as northern and southern Africa, the Mediterranean and central North America.

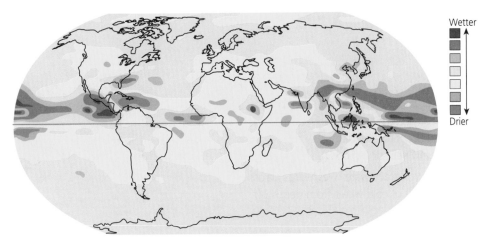

Wetter

Drier

Figure 1.3 Forecast global rainfall June–August if there is a doubling of atmospheric carbon dioxide

Factors driving the increase in natural disasters

Revised

Both physical and human factors are responsible for the increase in flood, tropical cyclone and drought disasters.

- Physical factors: global warming and El Niño-La Niña events.
- Human factors: population growth, urbanisation, poverty and the overexploitation of natural resources.

Global warming

There is conclusive evidence that the Earth's climate has warmed in the past 50 years (Figure 1.4). Globally, nine out of the ten warmest years on record occurred between 2000 and 2011.

However, it is debatable whether global warming is a natural process or the result of human activities (i.e. **anthropogenic**). We know that the global climate undergoes periodic shift because of astronomical cycles, changes in surface ocean currents and volcanic eruptions.

But the case for an anthropogenic cause of current global warming is strong. A close correlation exists between the rise in average global temperatures and the amount of carbon dioxide in the atmosphere. Before 1800, average carbon dioxide concentrations were around 270 ppm. Today they average 390 ppm and are rising rapidly. This is due largely to the burning of fossil fuels, although deforestation and the draining of wetlands have also played a part.

The **greenhouse effect** explains the link between global temperatures and carbon dioxide levels. **Greenhouse gases** (GHGs) such as water vapour, carbon dioxide and methane, which occur naturally in the atmosphere, absorb and re-radiate around 95% of the Earth's long-wave radiation. However, large increases in carbon dioxide and other GHGs during the past 200 years have led to more absorption of long-wave radiation by the atmosphere. The result is an **enhanced greenhouse effect** thought to be responsible for global warming and climate change.

Global warming has the potential to greatly modify the world's climate and cause huge disruption to human activities. But climate change is unpredictable and accurate forecasting of hazardous hydro-meteorological events presents enormous challenges.

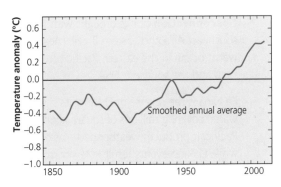

Figure 1.4 Global temperature change 1850–2010 (Source: Met Office (based on Brohan et al. 2006))

> **Typical mistake**
> There is no doubt that the Earth's climate is warming; the debate is whether this warming is due to natural processes or human activities.

> **Examiner's tip**
> A balanced explanation of the causes of global warming must include some reference to possible natural as well as human factors.

> **Typical mistake**
> The greenhouse effect is a natural phenomenon, caused by the presence of atmospheric CO_2, ozone and other gases, and warms the atmosphere. Global warming is an enhanced greenhouse effect, caused by human activities releasing large quantities of CO_2 and other gases into the atmosphere.

Now test yourself   Tested

6   Describe the possible link between global warming and the frequency of extreme rainfall events.

Answer on p. 121

El Niño and La Niña

El Niño and La Niña describe ocean temperature anomalies in the equatorial Pacific Ocean. These anomalies follow a cyclic pattern every 3 to 9 years, oscillating between warm (El Niño) and cold (La Niña) conditions — a phenomenon known as the El Niño Southern Oscillation (ENSO).

● In normal years the trade winds drive warm surface water westwards across the equatorial Pacific, leading to an upwelling of cold water from depth off the west coast of South America.

● During an El Niño event the trade winds weaken, and a layer of warm water in the western Pacific spreads eastwards.

● The result is unusually high sea surface temperatures (SSTs) in the central and eastern equatorial parts of the Pacific Ocean, and along the west coast of South America.

The Pacific Ocean covers nearly half the surface of the planet and therefore exerts a strong influence on global weather patterns. In El Niño years this **teleconnection** brings drought with forest and bush fires to Australia, southeast Asia and eastern Brazil. Meanwhile, enhanced evaporation from the warm ocean in the eastern Pacific often causes torrential rain and flooding in Peru and the southern USA.

La Niña is the reverse of El Niño. Its signature is an unusually cold tongue of water, covering millions of square kilometres, which develops in the equatorial Pacific. It produces weather patterns opposite to those of El Niño. For example, while El Niño events result in fewer tropical cyclones in the Atlantic, Caribbean, Gulf of Mexico and western Pacific, La Niña spawns more frequent tropical cyclones in these regions.

In recent decades higher ocean temperatures (due to global warming) have increased the intensity and frequency of ENSO. The outcome is likely to be less predictable and more extreme global weather.

> **Teleconnection** describes the relationship or coupling between surface ocean temperatures and regional climate and weather patterns.

> **Examiner's tip**
>
> Think of global warming as a giant experiment. Because the atmospheric system is so complex and has so many interconnections, the outcome of this experiment is unpredictable.

World population growth

The world's population almost quadrupled between 1900 and 2000, from 1,650 million to just over 6 billion. Although this phase of explosive growth is ending, growth continues: in 2012 global population passed 7 billion and by 2050 it will rise to around 9 billion. One effect of population growth is that millions more people are exposed to natural hazards. Also, most of this growth has been concentrated in the less economically developed world where populations are less able to cope with natural hazards and disasters.

Urbanisation

Rapid global **urbanisation** has occurred in the past 50 or 60 years. By 2012 over half the world's population lived in towns and cities.

Alongside urbanisation, **urban growth** has taken place at an unprecedented rate. Since 1950, the world's urban population has grown fivefold. More importantly, urban populations in LEDCs have increased by a factor of ten. Today, nearly 4 billion people live in towns and cities; 70% are in LEDCs and the vast majority are poor.

Now test yourself

7   What is El Niño?

Answer on p. 121
                              Tested

> **Urbanisation** is the proportion of urban dwellers in an area (world, country, region).
>
> **Urban growth** is the absolute increase in the number of urban dwellers.

Urbanisation and urban growth have increased disaster risks. People live at high densities in towns and cities, and hazard risks are further increased by poverty, poorly constructed buildings and a lack of **preparedness**.

Poverty

The World Bank's definition of poverty is an income of less than US$1.25 per person per day. Although the proportion of the global population living in poverty fell from 43% in 1990 to 22% in 2008, 1.3 billion people are still poor. Sixty-five per cent of them live in the world's least developed countries (LDCs). Thus nearly half of the population of Sub-Saharan Africa is poor, and this figure rises to over 80% in countries such as Liberia and Madagascar. South Asia also has high levels of poverty: in India, one-third of the population (nearly 400 million people) survive on less than US$1/day.

Poverty is a critical factor affecting people's vulnerability to natural hazards and natural disasters. At an individual level poor people often have little choice but to live in areas at high risk from natural flooding, landslides, earthquakes and other hazards.

Rich countries can afford to mitigate the effects of natural hazards, and thus reduce risk and vulnerability. Mitigating actions include the construction of levées, training, and coordinating emergency services and planning for immediate and longer-term recovery.

The overexploitation of natural resources

In some parts of the world the hazard risks have been increased by population pressure on natural resources. **Deforestation** in Indonesia was largely responsible for the sudden increase in floods, landslides and droughts between 2001 and 2009. In 1999, **debris flows** and **flash floods** in the coastal state of Vargas in northern Venezuela, which killed an estimated 30,000 people, were made worse by deforestation in the mountains adjacent to the coast (see case study on p. 12).

In Africa's Sahel region, population growth has forced indigenous farmers and herders to overcultivate arable land and overstock pastures. The resulting exploitation of natural resources is unsustainable and is evident in deforestation, disruption of the water cycle, more frequent droughts, **soil erosion** and **land degradation**. In southern Bangladesh, human pressure has destroyed coastal mangrove forests, increasing the threat of storm surges.

> **Examiner's tip**
>
> Balanced answers to questions on the increasing frequency and magnitude of natural disasters must consider a range of both physical and human factors.

> **Debris flows**, associated with extreme rainfall events and confined to river channels, comprise loose boulders, rocks, sand, mud, trees and other debris. They are fast-moving, powerful and have the consistency of wet concrete.
>
> **Flash floods** are high-flow events that develop quickly. They occur when high-intensity rain (e.g. a thunderstorm) in a river catchment induces rapid runoff.

> **Typical mistake**
>
> Millions of people, particularly in LEDCs, live in areas of high risk from natural hazards (e.g. floodplains, deltas). These people (often poor) are not unaware of the risks but have little choice if they are to access the soil, water and other resources needed to survive.

Case study — Hurricane Katrina

Hurricane Katrina hit the Gulf Coast of Louisiana at 06:10 on 29 August 2005. The storm killed 1,353 people and either destroyed or damaged 270,000 homes. Eighty per cent of New Orleans was flooded by an 8 m storm surge. Economic losses exceeded $US100 billion, making Katrina the costliest natural disaster in US history.

Causes. A combination of natural and human factors was responsible for the disaster.

- Many scientists believe that recent increases in the frequency and intensity of hurricanes in the Atlantic region are related to global warming. In 2005 sea surface temperatures in the Gulf of Mexico were 1°C above average, fuelling Hurricane Katrina.
- Rising sea level (also caused by global warming) puts lowland coasts along the Gulf of Mexico at increased risk from storm surges.
- Some 9.5 million people inhabit the coastal counties between Louisiana and Florida. In recent years population has grown rapidly along the coast, where densities are more than twice the US average.
- Land subsidence and the loss of wetlands in the Mississippi delta have increased exposure to storm surges. Subsidence is due to the extraction of natural gas. Reclamation of wetlands, which in the past stored flood water and absorbed the impact of storm surges, has weakened natural coastal defences. 'Walling-in' the Mississippi River with levées and the construction of dams upstream have starved the delta of the silt (from flooding) needed to keep pace with subsidence.
- The Mississippi Gulf Outlet, a 200 m wide canal linking New Orleans with the delta, acted as a funnel for the storm surge, increasing the surge's height by 20% and doubling its speed.
- Despite the known risks, the 560 km New Orleans' levée system had been poorly maintained and offered little protection against category 4 and 5 hurricanes.

Case study — Debris flows and floods in northern Venezuela

In December 1999, Vargas province on the northern coast of Venezuela was devastated by massive debris flows. Estimates of the final death toll ranged from 15,000 to 30,000. The economic impact was also severe: 20,000 houses were destroyed and 40,000 damaged. Hundreds of houses in Carmen de Uria were swept away, and large areas of Caraballeda, Macuto and Carmende were buried beneath flood debris.

Causes. The causes were meteorological, geomorphological and human.

- Between 8 and 19 December extreme rainfall in the coastal mountains (914 mm) triggered landslides, flash floods and debris flows.
- The Cordillera de la Costa (mountain range 2,000–2,700 m high) runs close to, and parallel with, the coast. Thus seaward-facing slopes are steep, runoff rapid and rivers have high energy.
- Igneous rocks that form the mountains have been deeply weathered and feed large amounts of sediment to streams and rivers that become debris flows.
- Widespread deforestation in the mountains further increased runoff and erosion.
- Because the mountains drop abruptly to the sea there is little flat land for settlement. Coastal alluvial fans, formed where rivers emerge from the mountains, offer some of the few sites for settlement. However, these locations are hazardous and exposed to floods and debris flows.
- In recent years rapid population growth has concentrated urban development on the high-risk alluvial fans.

Typical mistake

Explanations of the specific causes of natural disasters often lapse into general description/assessment of their impact and responses. Answers must remain focused and relevant to gain high marks.

Examiner's tip

Explanations of the impact of natural disasters should focus on the interaction between physical and human factors.

Natural disasters: mortality and economic trends

Revised

Mortality trends

Between 1970 and 2010 natural disasters accounted for some 3.3 million deaths worldwide. Average annual **mortality** was 82,500, with large year-to-year fluctuations. For example, in 2011 the Haiti earthquake alone killed around 170,000 people. However, when the number of deaths is compared with the growing world population, mortality caused by natural disasters has undergone a relative decline.

This downward mortality trend (which is a feature of both MEDCs and LEDCs) is largely explained by better preparedness and mitigation strategies (Figure 1.5) reducing vulnerability to tropical cyclones, floods, earthquakes and other natural hazards (Table 1.2).

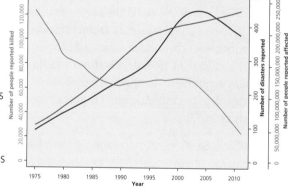

Figure 1.5 Natural disaster summary, 1975–2010 (Source: EM-DAT, www.emdat.be)

Table 1.2 Mitigating natural disasters

Early warning and monitoring

Tropical cyclones are closely monitored by weather satellites, aircraft, buoys and **radiosondes**. Early warnings give time for evacuation. Volcanoes are monitored by measuring ground inflation, gravity, earth tremors and gases released by **fumaroles**. Rainfall and river discharge are monitored to give early warning of flooding. Tsunami warning systems are in place in the Pacific and Indian Oceans. In the UK the Meteorological Office (Met Office) and Environment Agency issue warnings of extreme weather and floods. Advance warnings of tornadoes in the USA are given by the Storm Prediction Center.

Public education and planning

Public education helps to create societies that are more resistant to natural hazards. In Japan the public receive instruction on how to respond to earthquake and tsunami hazards. Disaster plans, designed by central government and local authorities, detail immediate and medium-term responses in the aftermath of major disasters.

Building codes

Many active seismic zones have building codes to prevent building collapse during major quakes, though in poorer countries these codes are often not enforced. Planning control may prevent the spread of residential and commercial development to areas of high risk, such as floodplains.

Radiosondes are balloons supporting an array of scientific instruments that measure the vertical distribution of temperature, pressure and humidity in the atmosphere.

Fumaroles are narrow vents on the slopes of volcanoes that emit steam and other hot gases.

Flood abatement is a type of flood management that aims to reduce peak discharge (through soft engineering) and therefore the risk of flooding.

Hard engineering
Levées, dams, sluice gates, etc. control river floods and storm surges. Storm shelters provide refuges against storm surge and tornado hazards.

Soft engineering
Successful approaches to **flood abatement** include reafforestation of catchments and restoring wetlands. Both reduce flood risks.

Economic trends

Disaster trends (excluding mortality), measured by the number of people affected and the economic costs, have shown significant increases in the past 50 years. In 1975, approximately 50 million people were affected by disasters; by 2010, the average number had risen to 240 million (Figure 1.5). Over the same period economic losses escalated from an annual average of $US10 billion to $US80 billion.

Several factors explain these trends:

● rapid world population growth, resulting in more people being exposed to natural hazards

● a growing proportion of the world's population concentrated in high-risk areas such as floodplains, **tectonic plate** boundaries, lowland coasts and deltas

● urbanisation: economic assets in towns and cities are much greater per unit area than in rural areas; potential economic damage from hazards is therefore greater and insurance claims much higher

● mitigation responses (e.g. levées, dams, flood insurance) induce a false sense of security and encourage development in high-risk locations.

Between 2000 and 2050 the World Bank estimates that urban populations exposed to tropical cyclone hazards or earthquakes will more than double. The implication is clear: the economic costs and number of people affected by natural disasters will continue to rise in future.

Typical mistake

Although the long-term trend of mortality caused by natural disasters is downwards, major disasters, such as the 2004 Asian tsunami and the Szechuan earthquake in 2008, result in great year-to-year variability.

Now test yourself

8 Why is the relative number of deaths caused by natural disasters decreasing?

9 Why is the economic cost of natural disasters increasing?

Answers on p. 121

Tested

Case study Northridge earthquake, Los Angeles 1994

Physical details and cause. The earthquake was 6.7 magnitude and centred on the suburb of Northridge, north of central Los Angeles (LA). It was caused by movement along a thrust fault connected with the San Andreas fault, and the northwest motion of the Pacific plate. The quake's focus was shallow: just 17 km.

Hazards. Violent ground shaking lasted for 10–20 seconds, damaging buildings and infrastructure. The quake triggered thousands of small landslides.

Exposure. The LA metropolitan area has a population of 16.5 million, an average density of 2,500 persons/km² and grew by nearly 60% between 1960 and 2000. The LA basin is one of the most seismically active areas in the USA. Exposure is therefore high.

Vulnerability. California is the richest state in the world's richest country. Massive investment has helped to reduce LA's vulnerability to earthquakes. The built environment is designed for seismic resistance. Stringent building codes and high levels of preparedness are in place.

Impact. The shallow earthquake focus and the density of buildings and infrastructure account for the massive damage caused by the quake. Although the death toll (57) was small, 680,000 people were affected, mainly through physical damage to residential properties and small businesses. The economic cost was massive — US$43 billion — making it the world's fourth most costly natural disaster up to 2012. Motorways collapsed at seven sites and 170 bridges sustained damage. Near the epicentre, well-engineered buildings survived the shaking without damage. Elsewhere structural failures pointed to deficiencies in design and construction. Many steel-framed buildings cracked and reinforced concrete columns were crushed. However, few buildings collapsed. Investigations following the quake showed a need to improve building codes.

Global hazard patterns

Assessing flood hazard risks Revised

Natural hazards, such as floods, and their impact in a local area can be assessed from historical, documentary and fieldwork evidence.

- There are nearly 800 gauging stations on British rivers. Data for average daily discharge, peak flows and **flow duration curves** are available for approximately the past 25 years from the Centre for Ecology and Hydrology (**www.ceh.ac.uk**).

- The Environment Agency website (**www.environment-agency.gov.uk/ homeandleisure/37837.aspx**) publishes flood risk maps for the UK at scales up to 1:10,000. The maps show areas at risk of flooding, the extent of extreme floods, flood defences and the areas that benefit from them.

- Newspapers (both local and national) provide information on the timing, scale and impact of flooding. They are especially useful for floods that pre-date gauged records.

- Fieldwork interviews with people affected by floods. Residents may provide information on flood dates, the depth of floodwater, damage to property, rates at which floodwaters rose, early warnings and flood insurance cover.

- Flood markers and plaques on buildings, walls and bridges. The scale of recent flood events can be investigated by observing the distribution of flood stones (i.e. cobbles and pebbles deposited by floods) and rubbish caught on overhanging branches above river channels.

> **Enquiry question**: Why are some places more hazardous and disaster-prone than others?

> **Flow duration curves** are graphs that show the percentage of time that a flow of given magnitude on a particular river is equalled or exceeded. Based on the historical record of flows, they are a useful predictor of flood risks.

> **Examiner's tip**
>
> Explanations of the impact of hazards and hazard risks should, wherever possible, be illustrated with actual examples. Ideally, examples should be local, which you have investigated personally.

Distribution of the world's major natural hazards

Revised

The global distribution of cyclone, drought, flood, earthquake, volcano and landslide hazards is a function of:

- the location of naturally occurring physical events such as floods and earthquakes
- the geographical distribution of population and economic activity

The interaction of these factors gives rise to hazard risks. Risks and the impact of hazards will depend specifically on the hazard's destructive potential (e.g. magnitude, duration, timing), exposure (number of people and economic assets in the area of the hazard), and the vulnerability of society (e.g. poverty, preparedness).

Cyclones

Cyclones are powerful revolving storms. They develop in the tropics and sub-tropics where they are known variously as tropical cyclones, hurricanes and typhoons. Extra-tropical cyclones, found in middle and high latitudes, are called **depressions**.

Tropical cyclones

Tropical cyclones create three natural hazards: high, **sustained winds** of more than $119\,km\,h^{-1}$, torrential rain and storm surges. Hurricane-force winds damage buildings, flatten trees and crops, and flying debris causes death and injury. Torrential rain creates river floods and triggers landslips, while storm surges cause flooding and widespread damage along lowland coasts. Tropical cyclones develop over the oceans between latitudes 8° and 20°. Three conditions favour their formation:

- high humidity and plentiful supplies of water vapour
- light winds that allow vertical cloud development
- sea surface temperatures of at least 26°–27°C.

Figure 1.6 shows that tropical cyclone hazards are concentrated on the eastern seaboards of landmasses, between latitudes 15° and 35°.

> **Examiner's tip**
>
> Note that tropical cyclone hazard risks are greatest in densely populated locations with large investments in economic infrastructure.

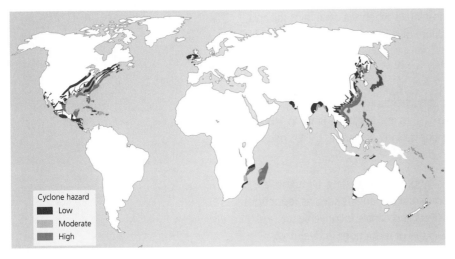

Figure 1.6 Global distribution of cyclone hazards (Source: World Bank)

Extra-tropical cyclones

Extra-tropical cyclones or depressions dominate weather and climate in middle and high latitudes. The main hazards brought by depressions are similar to tropical cyclones: strong winds (damaging **gusts** in exposed places may exceed 150 km h⁻¹); heavy rainfall leading to river floods and landslips; powerful waves, storm surges and coastal flooding, especially when low surface pressure and strong on-shore winds coincide with high tides.

Depressions form on the **polar front jet stream** which steers them from west to east in mid-latitudes. Hazard risks from depressions are high in northwest Europe because:

- for most of the year the jet stream, which generates the storms, lies close by
- of its position on the extreme western edge of the Eurasian landmass, exposed to 5000 km of open ocean, where winds are strong and storms are laden with moisture
- its high-density population, concentrated in towns and cities, increases economic risks

Droughts

Droughts are prolonged periods of abnormally low rainfall that adversely affect human well-being and economic activity.

Hazard distribution

Droughts are geographically more widespread than any other natural hazard (Figure 1.7). In western Europe, droughts are associated with persistent anticyclones that **block** the passage of depressions.

> The **polar front jet stream** is a narrow belt of fast-moving air that encircles the globe in mid-latitudes at a height of 8–10 km.

> **Typical mistake**
>
> You should understand the difference between the global distribution of natural physical events such as cyclones and earthquakes, and the global distribution of hazardous cyclones and earthquakes.

> **Now test yourself**
>
> 10 Name three natural hazards associated with tropical cyclones.
>
> **Answer on p. 121**
>
> Tested ☐

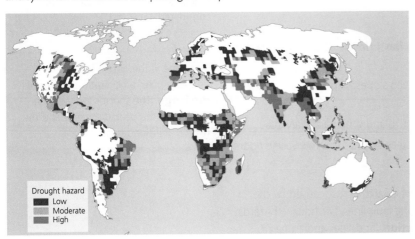

Figure 1.7 Global distribution of drought hazards (Source: World Bank)

Droughts also occur in the tropics and sub-tropics when seasonal rains fail and high pressure prevails. In south Asia the failure of the monsoon rains in 2009 led to drought and food shortages.

Hazard impact

Those parts of the world most susceptible to drought hazards are densely populated monsoon Asia, the Sahel, southern and eastern Africa and northeast Brazil. In MEDCs drought hazards are most problematic in continental interiors such as the US mid-west, Argentina's Pampas and Australia.

The impact of drought hazards in MEDCs is not life-threatening. Droughts are more likely to bring economic and environmental disbenefits such as reductions in crop yields, disruption of transport on rivers and canals, inconvenience (e.g. hose pipe bans) and damage to wildlife because of reductions in river flow.

River floods

River flood hazards occur when rivers overtop their banks, damaging property and infrastructure and sometimes causing death and injury.

Hazard distribution

Areas most at risk from flood hazards are densely populated (often highly urbanised) floodplains and deltas. Despite the risks, these areas are often attractive to human settlement because:

- valley floors provide flat land for building, are easily accessible by roads and railways and offer sites for river crossings
- floodplains and deltas comprise river-deposited sediments (**alluvium**) that support productive soils for agriculture
- river channels on floodplains and deltas are a source of water for irrigation, industry and domestic use

The highest rural population densities in the world are found on major floodplains such as the Ganges in northern India, the Indus in Pakistan, the Nile in Egypt and the Yangtze in China. Flood hazards increase where:

- rivers drain large catchments
- catchments receive heavy seasonal rainfall or snowfall
- there is little regulation of river flow (e.g. dams)
- deforestation accelerates runoff from catchments
- floodplains are densely populated and/or highly urbanised

All of these factors contributed to the Pakistan flood disaster in August and September 2010, when the Indus River and its tributaries flooded an area the size of England. The final death toll was approximately 2,000, with at least 1.2 million homes destroyed and whole villages swept away. Almost 10 million people were affected by the floods.

Many small river catchments are susceptible to **flash floods**. Flash flood hazards are greatest in catchments with:

- rapid runoff and short **lag times**, due to steep slopes, impermeable geology and sparse vegetation cover
- upland headwaters that intensify thunderstorms and other rainfall events
- settlements constrained in their location by steep slopes to narrow valleys floors

Earthquakes and volcanoes

Earthquakes are vibrations (**seismic waves**) in the Earth's crust caused by the fracturing of rocks and sudden movements along fault lines. Earthquake hazards include violent shaking of the ground, **liquefaction**, landslides and tsunamis. The precise location of an earthquake within the crust is known as the **focus**. The point on the surface immediately above the focus is the **epicentre**.

> **Typical mistake**
>
> The frequency and magnitude of river floods appear to be increasing. However, it is incorrect to attribute this trend solely to global warming and climate change. Population growth and rising pressure on environmental resources (deforestation, soil erosion, drainage of wetlands) must also be considered.

> **Lag time** is the difference in time between peak precipitation and **peak discharge**. The shorter the lag time the higher the peak discharge and the greater the flood risk.

Exam practice answers and quick quizzes at **www.hodderplus.co.uk/myrevisionnotes**

Volcanoes are openings in the Earth's crust where molten rock and gas reach the surface. Classic volcanoes have a conical shape. The **cone** comprises layers of lava, ash and other ejecta erupted at the surface. The vent occupies a collapsed hollow that, depending on its size, is known as a **crater** or **caldera**. Feeding the volcano, and located 3–4 km below the crater is the **magma chamber**. Magma also erupts from rifts or **fissures** — a type of eruption that is common in Iceland and Hawaii.

The global distribution of earthquakes and volcanoes

The world's major seismic zones correspond to active **tectonic plate** boundaries. At these boundaries **inter-plate movements** cause tension, compression and shearing of the rocks. Rocks under pressure eventually snap, resulting in crustal movements along fault lines or earthquakes. Major earthquakes can also occur thousands of kilometres away from plate boundaries. They are known as **intra-plate quakes**.

Most volcanoes are clustered around destructive and constructive plate margins where magma reaches the surface. The only exceptions are volcanoes located at hotspots such as Hawaii.

Plate tectonics

The Earth's outermost layer or **lithosphere** is broken into seven large slabs and a dozen or so smaller ones known as **tectonic plates**. The global distribution of the main tectonic plates is shown in Figure 1.8. Driven by convection currents deep in the Earth's interior, tectonic plates are in constant movement. This dynamism, which is most apparent at plate margins, is responsible for most earthquakes and volcanic eruptions.

> The crust and the immediate underlying rigid part of the mantle form a single unit known as the **lithosphere**. Tectonic plates consist of slabs of lithosphere.

> ### Now test yourself
>
> 11 What is the difference between an earthquake focus and an earthquake epicentre?
>
> **Answer on p. 121**
>
> Tested

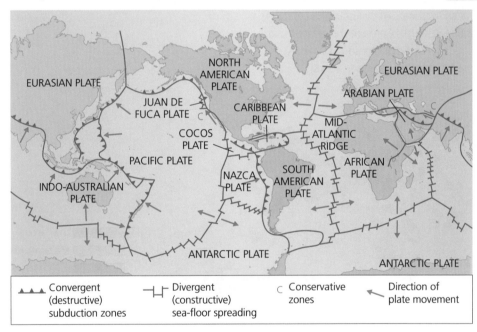

Figure 1.8 Distribution of the main tectonic plates

There are three types of plate margin: **destructive (subduction zones)**, **constructive** and **conservative** margins. Each has its own distinctive tectonic processes.

Tectonic activity at destructive plate margins.

- Oceanic crust/lithosphere is destroyed. Subduction occurs when two tectonic plates converge. The older, denser plate is driven down into the upper mantle (i.e. it is subducted) where, at depth, it melts and is destroyed.
- Earthquakes are common. They are caused by frictional resistance and compression between the subducted plate and the upper mantle.

> **Typical mistake**
>
> It is wrong to assume that earthquakes and volcanic activity are confined only to tectonic plate margins. Many earthquakes occur along fault lines thousands of kilometres from plate boundaries. Volcanic activity also occurs outside plate boundary zones, e.g. Hawaii, Canary Islands.

● Volcanic activity due to melting of the subducted plate and the surrounding mantle rocks. This melt, which is often **viscous** and gaseous, slowly rises to the surface. Where it breaks the surface it forms volcanoes that often erupt explosively (e.g. Mount St Helens, Krakatoa).

Constructive plate margins. At constructive plate margins rising plumes of magma from the upper mantle stretch the crust and lithosphere. The resulting drop in pressure causes melting and volcanic eruptions on the ocean floor. Meanwhile, parallel faulting caused by tension in the crust is the focus of shallow earthquakes and leads to the formation of submarine **rift valleys** that extend for thousands of kilometres along the ocean floor.

Conservative plate margins. At conservative plate margins (e.g. coastal California) tectonic plates slide slowly past each other with a shearing movement. This movement is rarely smooth. If the plates lock and pressure builds, sudden movements may generate powerful earthquakes.

Volcanic hotspots. These are places where a plume of magma rises from the mantle, punches a hole through a tectonic plate, and erupts at the surface. Hotspots are sites of intense volcanic activity and effusive eruptions of basaltic lava, e.g. Hawaii.

> **Examiner's tip**
>
> Familiarity with the processes that operate at tectonic plate margins will help you understand the nature of geophysical hazards such as earthquakes and volcanic eruptions.

> **Examiner's tip**
>
> Learn to draw simple annotated cross-sections to show the processes and landforms at destructive and constructive plate boundaries. Remember that a picture can often say more than a thousand words.

Earthquake hazards

Twenty per cent of the world's population lives in seismically active zones. The global distribution of earthquake hazards is concentrated in these areas (Figure 1.9). Seismic zones that generate frequent high-magnitude quakes and have high-density urban and rural populations encircle the Pacific Ocean (i.e. the 'Pacific Ring of Fire'). Earthquake hazard risks are also high in the eastern Mediterranean (Turkey), the Middle East (Iran), northern India, Pakistan (Kashmir), Afghanistan and central Asia. Tsunami hazards, triggered by submarine earthquakes, present the greatest risks where densely populated coastal regions (e.g. northeast Honshu in Japan, Aceh province in Indonesia) are situated close to offshore subduction zones.

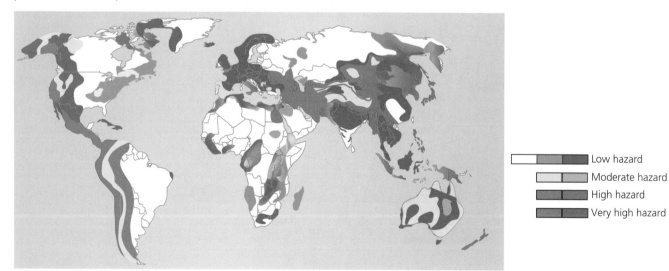

Low hazard
Moderate hazard
High hazard
Very high hazard

Figure 1.9 Global seismic hazards

Volcanic hazards

Volcanic eruptions produce a range of hazards (Table 1.3). Their impact on people depends on:

● the nature of the volcanic ejecta (i.e. lava, tephra, gas)

● the density of population and settlement in the vicinity of the eruption

● the preparedness of the population at risk

Gentle, **effusive eruptions** of lava pose little direct threat to human life, though lava flows may cause extensive damage to property and infrastructure. On the other hand, **explosive eruptions** of superheated gas and tephra (pyroclastic flows) cause total devastation.

Table 1.3 Volcanic hazards

Lava flows	Lava flows can cause enormous damage to property, although they are rarely a threat to human life. On Lanzarote lava flows in 1730–1736 buried most the island's farmland.
Pyroclastic flows	Pyroclastic flows are high-speed avalanches of hot ash, rock fragments and gas that destroy everything in their path. A pyroclastic flow on Martinique in 1902 killed 30,000 people — the island's entire population.
Lahars	Lahars are mixtures of water, rock, sand and mud that flow down valleys leading away from a volcano. They are fast-moving and can travel long distances.
Jökulhlaups	Eruptions that occur beneath an icefield or glacier cause rapid melting and floods known in Iceland as jökulhlaups. The eruption of Grímsvötn beneath the Vatnajökull icefield in southern Iceland in 1996 triggered a massive jökulhlaup that destroyed several bridges and 10 km of the ring road that encircles the island.
Ashfalls and tephra	Tephra comprises fragments of volcanic rock blasted into the atmosphere by explosive eruptions. Fall-out of tephra is highly disruptive. It blankets the landscape in ash, houses collapse and jet engines malfunction. The Eyjafjallajökull eruption in Iceland in 2010 brought European air traffic to a standstill for 6 days.

Although volcanic eruptions can release ash high into the stratosphere, temporarily modifying global weather and climate, the impact of lava flows, pyroclastic flows and lahars is mainly felt within 50 km of the site of eruption. Volcanic hazards present the greatest threat where high-density urban areas are close to active volcanoes, e.g. Naples (close to Vesuvius) and Mexico City (close to Popocatepetl). (See Figure 1.10.)

Landslides and avalanches

Driven by gravity, **mass movements** such as landslides and avalanches transfer slope materials such as rock, soil and snow downhill as a coherent body. Landslides and avalanches occur rapidly and often without warning. This makes them extremely hazardous.

The causes of mass movement

Mass movements indicate slope failure. They occur when forces operating downslope (e.g. gravity, mass of slope materials) exceed resisting forces (e.g. the frictional resistance between slope materials and shear strength of materials).

Mass movement may be triggered by:

- steepening and undercutting at the base of a slope, either by natural erosion (e.g. a valley-side slope undercut by a river) or human activity (e.g. a road cutting)
- heavy rainfall (i.e. loading), which adds to the mass of material on the slope, lubricates slope materials and reduces their coherence
- deforestation removing the binding effect of tree roots on slopes
- earthquakes reducing the resistance of slope materials through violent shaking

Distribution of landslide and avalanche hazards

Globally, around 3.7 million km² are susceptible to landslide and avalanche hazards. The global distribution of these hazards is closely associated with two types of region (Figure 1.10): (a) major fold mountain ranges such as the Himalayas and the Andes, and (b) regions of seismic activity and volcanism (e.g. Japan, SE Asia).

Examiner's tip

Landslides, avalanches, earthquakes and volcanic hazards often occur in geographical proximity because of their common link to tectonic plate boundaries.

Examiner's tip

Landslides and avalanches occur when changes such as slope erosion or heavy snowfall create disequilibrium in the balance of forces operating on slopes.

Figure 1.10 Global distribution of landslide hazards (Source: World Bank)

Now test yourself | Tested

12 Which of the following natural hazards often occur at the same time as landslides: cyclones, earthquakes, coastal floods, droughts, volcanic eruptions?
13 Which natural hazard affects most people in Africa?

Answers on p. 121

Disaster hotspots — Revised

Disaster hotspots are geographic areas where multiple hazard risks are found (Figure 1.11). The California coast and the Philippines are disaster hotspots at opposite ends of the development spectrum.

Now test yourself

14 What is a 'disaster hotspot'?

Answer on p. 121

Tested

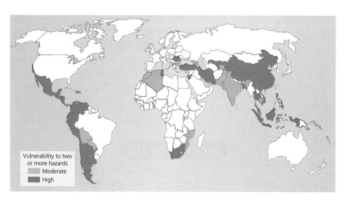

Figure 1.11 Disaster hotspot countries (Source: World Bank)

Case study The California coast

The California coast between San Francisco and San Diego is a disaster hotspot. In the past 25 years this region has experienced two major earthquakes, severe droughts, wildfires and river floods. Extreme natural events are a feature of California's geophysical environment and its climate. Despite a high state of preparedness and effective disaster planning, natural hazards often become disasters because of California's:

● large urban population (more than 20 million)
● huge and costly investment in real estate and social and economic infrastructure in the coastal belt

Earthquakes

California is one of the most seismically active regions in the world. Crustal instability and earthquakes are caused by movement along the San Andreas fault (and scores of hidden thrust faults), which separates the Pacific and North American tectonic plates.

During the twentieth century the San Andreas fault triggered major earthquakes (magnitude 6.7–7.8) in San Francisco in 1906, Loma Prieta in 1989 and Northridge in 1994.

San Francisco earthquake 1906. The 1906 earthquake was a major disaster for the city of San Francisco. A magnitude 7.8 quake shook the ground for nearly 1 minute, resulting in damage estimated at US$80 million. However, the damage caused by fires which swept through the city was five times greater. Overall the quake destroyed 80% of San Francisco. At least half the population was made homeless; fatalities numbered between 3,000 and 6,000.

Damage caused by the quake was particularly severe because: (a) most buildings were wooden structures, susceptible to fire and (b) much of the city was built on land reclaimed from San Francisco Bay. This reclaimed land, mainly comprising water-filled sediment dredged from the Bay, underwent **liquefaction** during the quake. As a result, thousands of buildings collapsed.

Drought

California experienced three successive years of drought from 2007 to 2009. The drought had severe impacts on California's economy, while water shortages, together with high summer temperatures and strong winds, caused devastating wildfires. Drought was due to a persistent blocking anticyclone over the southwest USA and the North Pacific Ocean, forcing the jet stream north of its usual track.

The 2007–2009 drought had major economic impacts, particularly on agriculture. By 2009 water allocations for irrigation in the western Central Valley had been reduced by 90%. Many farmers had to leave arable land fallow, while avocado and citrus growers were forced to cut down orchard trees because of insufficient irrigation water. Total agricultural losses for 2009 were reported as $US700 million and 21,000 jobs in the farm sector were lost.

Hydroelectric power (HEP) generation, fisheries and wildlife, and recreation were also badly affected by the drought. In 2008 low water levels in rivers and reservoirs reduced the proportion of electrical energy generated by HEP in California to just 8%, compared with 16.6% in 2006. Low river levels were also responsible for the death of fish and disruptions to the migration of Pacific salmon. Lack of snow in the Sierra Nevada affected ski resorts, and low water levels in reservoirs and rivers limited recreational activities such as sailing and rafting.

Wildfires

Causes. Wildfires are a major hazard in southern California. Risks are greatest in late summer and autumn and especially during periods of drought and prolonged hot weather. Hot, dry summers desiccate the vegetation, creating tinder-box conditions and abundant fuel for wildfires. Adding to the risk are the local Santa Ana winds, which blow in autumn and early winter. They are hot, dry and gusty; they fan the flames of wildfires and spread embers.

The frequency and intensity of wildfires increased during the 2007–2009 drought. Five of the ten biggest wildfires ever recorded in California occurred during this three-year period.

In 2008 wildfires in California burned approximately 525,000 ha of land and destroyed more than 1,000 structures. The total economic cost was nearly $US900 million. (Losses were even greater in 2007 when 20% more land was burned and nearly 4,000 structures were destroyed.) In May 2009 a wildfire in Santa Barbara County raged for 15 days, destroying 3,500 ha of forest and 80 homes. At its height it even threatened the city of Santa Barbara and one-third of its 90,000 residents were evacuated.

> **Liquefaction** is the process that causes soils, saturated with water, to lose their strength and cohesiveness when subject to the stress of ground shaking in an earthquake. In this situation soils start to behave like a liquid.

Typical mistake

The magnitude of an earthquake is only one of the physical factors that affect its impact. The location of major centres of population in relation to the epicentre and the depth of the earthquake's focus are also important.

Examiner's tip

You should be aware that many natural hazards are interconnected. For example droughts may be linked to famine and wildfire hazards, and earthquakes are a frequent cause of landslide hazards.

Case study The California coast

The Philippines is a lower-middle income country. Situated in the western Pacific between latitudes 5° and 20°N, it forms an archipelago of hundreds of islands. The largest islands are Luzon and Mindanao. The Philippines is a disaster hotspot because:

- located on the small Philippine plate, the country is surrounded by active plate boundaries and subduction zones
- there are approximately 25 active volcanoes, and eruptions and earthquakes occur frequently
- steep slopes, heavy rainfall and mountainous terrain promote mass movements
- extreme rainfall caused by tropical cyclones and the southwest monsoon are a feature of the climate
- the country has a large population (103 million in 2012) and a high average population density (300 persons/km²)

Volcanic eruption: Mount Pinatubo

After lying dormant for 500 years, Mount Pinatubo on the island of Luzon erupted in June 1991. The eruption was the second biggest of the twentieth century. Its cause was the subduction of the Eurasian plate beneath the Philippine plate along the destructive plate margin to the west of Luzon (Figure 1.12).

The eruption pumped volcanic ash clouds high into the stratosphere and was accompanied by pyroclastic flows and lahars. Thousands of houses collapsed under the weight of the ash, which also blanketed farmland. The combination of ash and sulphur dioxide released in the eruption lowered average global temperatures by 0.5°C between 1991 and 1993.

Without vegetation cover, ash deposits were easily mobilised by heavy monsoon rains and formed giant lahars. They buried more than 100 km² of farmland in the lowlands and destroyed everything (including several villages and towns) in their paths. Lahars continued to be a hazard for six years after the eruption.

Around 1 million people lived around the volcano. A total of 20,000 indigenous people living on the slopes of Mount Pinatubo were permanently displaced, and 200,000 people in the lowlands evacuated. The final death toll was nearly 800.

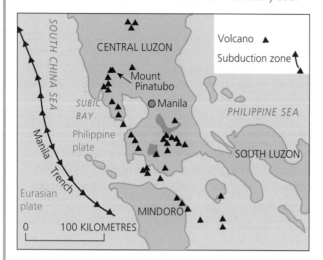

Figure 1.12 The Philippines: tectonic plates and volcanoes

The Guinsaugon mudslide

A devastating mudslide hit the Philippine village of Guinsaugon in southern Leyte province on 17 February 2006 (Figure 1.13). The slide covered 9 km², was 3 km wide and in places 30 m thick.

The main cause of the disaster was extreme rainfall associated with a La Niña event in the western Pacific. In 10 days, 200 cm of rain fell, loading slopes and weakening slope materials. There were, however, other contributory causes.

- Slopes in the region are steep and the volcanic rock is deeply weathered, making landslides, mudslides and other mass movements common.
- Widespread deforestation during the past 70 years (because of population growth and commercial, often illegal, logging) has increased slope instability.
- A small earthquake measuring 2.6 m, occurred immediately before the mudslide.

- Villages sited at the foot of steep slopes, occupied the run-out zone of the mudslide.
- Rapid population growth (2.7% per year 1995–2000) forced people to live in areas at high risk from landslide and mudslide hazards.

Impact. The mudslide moved at speed, burying the village of Guinsaugon. People had no warning and no time to escape. Survivors described how a 'wall of mud' descended on the village, killing more than 1,000 people, including 246 children at the local primary school. Virtually every one of the 300 houses in Guinsaugon was destroyed. The slide also killed thousands of livestock and buried surrounding farmland. Altogether around 16,000 people were affected.

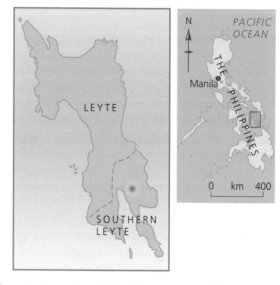

Figure 1.13 Location of the Guinsaugon mudslide

Examiner's tip

Case studies must include place-specific detail and not rely on generalisations that could apply to many different places. Examples of specific detail include locations and place names, dates, local details of geology, populations, magnitude of hazard events, etc.

Examiner's summary

Global hazards

✔ Extended-answer questions on natural hazards and disasters should be illustrated with contrasting examples and case studies.

✔ Effective case studies of hazard and disaster events must provide clear and specific details of places, timings, magnitude, impact, etc.

✔ Where appropriate, descriptions and explanations of physical processes should be supplemented with annotated sketch maps and diagrams.

✔ Extreme natural events such as earthquakes and droughts become natural hazards when they adversely affect people.

✔ Debate on the issue of global warming centres not on its reality, but whether its causes are natural or anthropogenic.

✔ A balanced discussion of global warming and climate change will consider a range of natural and human causes.

✔ Exam answers must focus exclusively on the question set. For example, discussions on the causes of natural disasters must not consider their impact and mitigation.

✔ Because of poverty, millions of people are forced to live in places where they are at risk from natural disasters.

✔ Although the global mortality trend of natural disasters is down, this trend masks considerable year-to-year variability.

✔ Explanations of earthquake, volcanic and other hazards and disasters should be underpinned by a sound understanding of physical processes.

Climate change and its causes

Timescales for climate change

Long-term and medium-term climate change

Over the past five or six decades average global temperatures have increased by approximately 0.5°C, a trend popularly known as global warming. However, this warming must be seen in the context of natural climate change occurring over much longer timescales. Thus the climate record over the past 500,000 years shows the Earth's climate to be highly unstable, with dramatic fluctuations in global temperatures (Figure 1.14).

During this time there have been four major glacial cycles with cold **glacial** periods (ice ages) followed by warmer **interglacials**. Each cycle lasted around 100,000 years. At the height of the last glacial 20,000 years ago, average annual temperatures in Britain were 5°C lower than today, and Scotland, Wales and most of northern England were submerged by ice up to 1 km thick. In the warm interglacial periods, average global temperatures were similar to those of today. Currently the world is in one of these warmer phases — the Holocene — which has lasted for 10,000 years.

On geological timescales, global temperature changes have been even more extreme. For example, 250 million years ago, at the end of the Permian period, average global temperatures reached 22°C — at least 7–8°C warmer than today. Similar temperatures prevailed throughout the Cretaceous period, and for much of the Tertiary era.

> **Enquiry question**: Is global warming a recent short-term phenomenon or should it be seen as part of longer-term climate change?

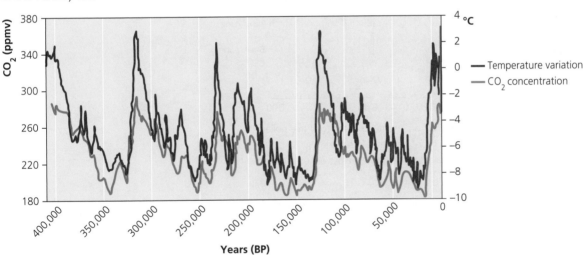

Figure 1.14 Global temperatures and CO_2 concentrations, 450,000 BP to present

Now test yourself

Tested ☐

15 What are glacial and interglacial periods?

Answer on p. 121

> **Examiner's tip**
>
> It is important to put present-day global warming in context. The climate record for the medium and long term tells us that the Earth's climate is unstable and has undergone huge fluctuations in the past.

Evidence for long- and medium-term climate change

Evidence for climate change comes from sea-floor sediments, ice cores, pollen analysis, dendrochronology and historical records. The forensic evidence is described in Table 1.4.

Table 1.4 Forensic evidence for climate change

Sea-floor sediments	The fossil shells of tiny sea creatures called **foraminifera** that accumulate in sea-floor sediments can be used to reconstruct past climates. The chemical composition of foraminifera shells indicates the ocean temperatures in which they formed.
Ice cores	Ice cores from the polar regions contain tiny bubbles of air — records of the gaseous composition of the atmosphere in the past. Scientists can measure the frequency of hydrogen and oxygen atoms with stable isotopes. The colder the climate, the lower the frequency of these isotopes.
Pollen analysis	Pollen analysis allows scientists to reconstruct past vegetational changes. From these data palaeoclimatic conditions can be inferred. Pollen diagrams (Figure 1.15) show the number of identified pollen types in the different sediment layers either as a direct count or as a percentage of total tree pollen.
Dendrochronology	Dendrochronology is the dating of past events such as climate change through study of tree-ring (annule) growth. Annules vary in width each year depending on temperature and moisture availability.
Retreating glaciers	Glacial erosional landforms such as cirques and U-shaped valleys in Britain's uplands, and lowland depositional features such as drumlins and moraines, are evidence of past glaciations. Glaciers returned briefly to the uplands during a brief cold snap known as the Loch Lomond Stadial (12,500–11,500 BP), leaving behind terminal, lateral and hummocky moraines that remain fresh in the contemporary landscape. Today 98% of the world's valley glaciers are retreating.

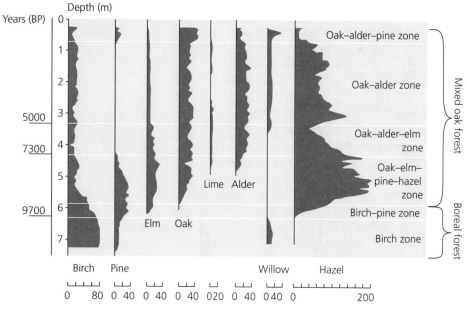

Pollen (except hazel) is expressed as a percentage of the total tree pollen.
The hazel scale relates to pollen count per unit of peat.
Age is shown on the left; equivalent modern forest types are on the right.

Figure 1.15 Tree pollen counts

Historical records also provide information about past climates. Europe experienced a so-called 'Little Ice Age' between the mid-fourteenth and early-nineteenth centuries. Seventeenth-century diarist John Evelyn described frost fairs on the Thames in London, when the river froze. Such events were captured by contemporary artists such as Abraham Hondius. The sixteenth-century Dutch artist Pieter Breugel the Elder is famous for his winter landscapes, with snow, frozen lakes and skaters. In 1783 many scientists and naturalists (including Gilbert White and Benjamin Franklin) recorded the effects of the Laki volcanic eruption in Iceland in Europe and North America. They described ashfalls and a thick sulphurous smog that settled over much of the Northern Hemisphere. They also noted the cooling effect the eruption had on the summer of 1783, and the exceptionally severe winter that followed.

Evidence of short-term climate change

Compelling evidence for climate change over the past three or four decades is available from meteorological, ecological and environmental sources.

Meteorological

The consistent rise in average global temperatures since the mid-twentieth century and forecasts for further increases of 2–3°C by the end of the present century,

Now test yourself

16 What landscape evidence in Britain's uplands demonstrates that the climate has changed dramatically in the past 20,000 years?

17 What was the 'Little Ice Age'?

Answers on p. 121

Tested

provide strong evidence for climate change. Moreover, rates of global temperature increase appear to be accelerating: nine out of the ten warmest years on record occurred between 2001 and 2011. Compared with the early 1970s, in the UK spring now arrives 6 days earlier and autumn 3 days later.

Ecological

Climate change puts pressure on natural ecosystems. As habitats change, some plant and animal species adapt by migrating either latitudinally or altitudinally. In the UK, global warming has provided opportunities for species such as little egrets, Cetti's warbler and tongue orchids to colonise. Some British songbirds that previously spent winter in southern Europe (e.g. chiffchaffs, blackcaps) now overwinter in southern England. Surface temperatures in UK coastal waters have increase by 0.7°C in the past 30 years. As a result, species such as basking shark, swordfish, tuna and sunfish have become more common. Meanwhile cold-water species such as cod and haddock have moved further north. In the tropics, rising sea temperatures have caused **coral bleaching**. Some stag head corals, unable to adapt, have shown significant decline.

Environmental

Most of the world's **glaciers** have been losing mass for at least the past 40 years. Rates of melting have increased since the 1990s and glaciers in the European Alps may shrink by 80–96% by 2100. Melting of the Greenland and West Antarctic ice sheets since the early 1990s has also been rapid. In 2002 the Larsen B **ice shelf** in Antarctica, covering 3250 km^2, broke up in less than a month. One-fifth of Antarctica's **sea ice** melted between 1950 and 2000. Melting and thinning of Arctic sea ice is even more alarming. At the end of August 2012 Arctic sea ice for the first time covered an area of less than 4 million km^2 — 3 million km^2 below the seasonal average for 1979–2000. The thickness of sea ice is estimated to have halved since 1979 and by 2030, the Arctic Ocean could be completely ice-free in summer.

Melting of glaciers and ice sheets, together with the thermal expansion of the oceans, are responsible for rising sea levels. Sea level rose by 20 cm between 1900 and 2000. Computer models suggest an average rise of sea level by at least 1 m by the end of the century.

> **Typical mistake**
>
> Because of the complexity of the global climate system, trends in global warming will always show some exceptions. Thus in the Karakoram Range in the western Himalayas glaciers are currently advancing (because of increased snowfall). In April 2012 Arctic sea ice was at its most extensive since April 2001.

The causes of climate change
Revised ☐

Natural causes

Astronomical events

Long-term climate shifts such as glacial cycles are caused by astronomical events. Glacial cycles are driven by changes to the Earth's axis and orbit, which affect the amount of solar radiation reaching the planet's surface.

- Today the Earth is tilted on its rotational axis at an angle of 23.4° to its orbital plane. But over a period of 41,000 years the angle of inclination fluctuates between 22° and 24.5°.
- The eccentricity of the Earth's orbit around the Sun varies from near-circular to elliptical. The two extremes are separated by 96,000 years.
- The Earth gyrates like a spinning top on its axis: this phenomenon, known as the **precession of the equinoxes**, affects the intensity of the seasons.

Together these astronomical changes combine to create alternating glacial and interglacial periods known as **Milankovitch cycles**.

Ocean currents

Shifts in surface ocean currents can cause sudden climate change. Around 12,000 years ago the warm North Atlantic current shut down, and brought ice-age

> **Now test yourself**
>
> 18 Name three astronomical causes of climate change.
>
> **Answer on p. 121**
>
> Tested ☐

conditions to northwest Europe (the so-called **Loch Lomond stadial**). The resulting **cool down**, which lasted for nearly a millennium, caused glaciers to return to Scotland, the Lake District and Wales.

Volcanic eruptions

Major volcanic eruptions can cool the global climate for several years, and in theory could trigger glaciations. Large-scale eruptions pump huge amounts of volcanic ash and sulphur dioxide into the stratosphere, reducing insolation at the Earth's surface. Mount Pinatubo, which erupted in June 1991, cooled the global climate for a couple of years, briefly halting global warming.

Anthropogenic causes

Global warming

There is conclusive evidence that Earth's climate has warmed in the past 150 years. For example:

- nine out of the ten warmest years on record occurred between 2001 and 2011
- average global temperatures increased by 0.4°C between 1992 and 2010
- the highest-ever temperature recorded (38.5°C) in the UK was in August 2003, the highest July temperature (36.1°C) was recorded in 2006, and maximum temperatures in March 2012 were 3.6° above the average — a record for the month

Today, the overwhelming view of scientists is that global warming is a reality. But, there is still debate on its cause. Is global warming a natural trend or is it due to human activities?

The enhanced greenhouse effect

The **greenhouse effect** explains the link between global temperatures and CO_2 levels. Greenhouse gases (GHGs) such as water vapour, carbon dioxide and methane (CH_4) occur naturally in the atmosphere. They absorb and re-radiate around 95% of long-wave radiation emitted by the Earth. CO_2 alone raises global temperatures by 7°C. Without this natural greenhouse effect average global temperatures would be 30°C lower. However, large increases in CO_2 and other GHGs during the past two centuries have led to more long-wave radiation being aborbed by the atmosphere. The result is an **enhanced greenhouse effect** driving global warming and climate change.

The **enhanced greenhouse effect** amplifies the Earth's natural greenhouse process as human activity releases additional CO_2 and other GHGs to the atmosphere.

Now test yourself

19 What is the enhanced greenhouse effect?

Answer on p. 121

Tested

Recent climate change

Revised

Most climatologists believe that the climate changes observed over the past half century are different from the past. First, changes are happening much faster than any previous event. And second, there is strong evidence linking these changes to human activities (Figure 1.16).

The case for an anthropogenic global warming rests primarily on the correlation between the rise in average global temperatures during the past 60 years, and the increase in CO_2 in the atmosphere over this period. Before 1800 average atmospheric CO_2 concentrations were around 270 ppm. In 2012 the figure was 387 ppm and rising rapidly. This increase is due largely to human activities — in particular massive growth in the consumption of fossil fuels. Other anthropogenic factors implicated in rising CO_2 and other GHG levels include deforestation, the draining of wetlands, permafrost melting, and intensive livestock farming.

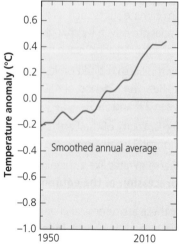

Figure 1.16 Global average temperatures 1950–2010 (Source: Met Office (based on Brohan et al., 2006))

The impacts of global warming

The direct impacts

Revised

Climate change is already having direct environmental, ecological and economic impacts on fragile ecosystems in the Arctic, and in many parts of Sub-Saharan Africa.

> **Enquiry question**: What are the impacts of climate change and why should we be concerned?

Case study — The impact of climate change in Alaska and Arctic Canada

Natural ecosystem. In Alaska and Arctic Canada low-growing woody plants, grasses, mosses and lichens dominate the tundra ecosystem. Average temperatures are below 0°C for up to 9 months and the growing season is short. Apart from the surface layers, which thaw in summer, the ground remains permanently frozen (**permafrost**). The harsh environment results in low productivity, low biodiversity, small animal populations and short **food chains**.

Climate change. Global warming is already bringing profound change to the Arctic. Large temperature increases (1°C per decade) have been recorded since 1970. This is due to decreases in **albedo** (surface reflectivity) as snow, ice and sea-ice melt, causing the ground and sea to absorb more solar radiation. The poleward advance of trees and shrubs will further lower albedos and accelerate warming.

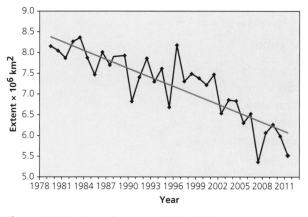

Figure 1.17 Arctic sea ice extent 1979–2011
(Source: National Snow and Ice Data Center)

Ecological and environmental impact. Many native species, intolerant to warmer conditions, will become extinct. As the treeline advances, open tundra habitats will shrink and give way to coniferous forest. By the end of the century up to 90% of Alaska's tundra may disappear. These changes will have drastic effects on **food webs** and wildlife. For example, melting of sea ice (Arctic Sea ice cover was at a record low in 2012) will prevent seals hauling out to give birth to their young (Figure 1.17) and polar bears, which depend on the sea ice to hunt for seals and other marine mammals, will starve. Warmer conditions will see more insects and pathogens spreading plant diseases and, as the forest advances, migrant birds will lose their summer wetland breeding grounds.

Global warming is already melting the permafrost, releasing methane, carbon dioxide, and other GHGs. Because permafrost stores so much carbon dioxide and methane is such a potent GHG, permafrost melting could massively accelerate global climate change in future.

Economic impact. Climate change in the Arctic will impact heavily on indigenous people such as Inuit Inupiaq and the Gwich'in tribes of northern Alaska. The Inupiaq's economy, based on hunting whales, seals and walruses, is threatened by the melting of sea ice. The Gwich'in or 'people of the caribou' depend almost entirely on the annual migration of the caribou herds. Any change to migration patterns will destroy a way of life and culture that has existed for millennia.

Climate change could also affect commercial operations in the Arctic. Exploration for oil and gas can only take place in winter when the ground is frozen. Permafrost melting will mean that the ground is unable to support heavy machinery. The physical stability of buildings and transport infrastructure (roads, pipelines) also depends on conserving the permafrost.

Examiner's tip

Learning case studies is essential preparation for the AS exam. Remember that some case studies are prescriptive. This means that exam questions can refer to particular case studies listed in the specification, e.g. Arctic region, Africa, California, Philippines.

Case study — Economic impact of global warming in Africa

The economic effects of global warming will be more severe in Africa than in any other major region. Climate change is already having an economic impact and has become a key development issue in Africa.

Climate change. By 2080, average temperatures in Africa could be 2°–3°C higher. In addition, the climate will be more variable, with greater extremes of temperature and rain. Drought, which 50 years ago in East Africa occurred on average once a decade, now occurs every two or three years.

Africa's vulnerability. Africa, the world's poorest continent, is least able to cope with climate change. Apart from poverty, it suffers more than any other region from corrupt governance, civil wars and tribal conflicts. It also relies heavily on agriculture (70% of all employment), which is highly sensitive to climate change. Only 4% of African farmland is irrigated, making agriculture susceptible to drought.

Agriculture, food supplies and environment. Crop yields could fall by 16% by 2080. This has obvious implications for

food security. Climate change also increases competition and conflict for scarce resources such as land, water, pasture and forests. In the Sahel in West Africa, population pressure has led to uncontrolled exploitation of natural ecosystems, resulting in **land degradation**, **desertification**, poverty and the collapse of entire farming communities.

Famine in the Horn of Africa. In 2011 **famine** hit parts of Somalia, Kenya, Djibouti, Ethiopia and Uganda. UNICEF claimed than 'tens of thousands' died and that 3.7 million required humanitarian aid. In the worst-hit areas **malnutrition** affected up to half the population. Famine was caused by (a) a 2-year drought, which killed up to 90% of livestock and resulted in record food price inflation, and (b) civil war and decades of corrupt governance. Recent droughts and floods have reduced GDP in Ethiopia by one-third, and in Kenya by 10–16%.

Disease. More frequent droughts and food shortages will increase the prevalence of diseases such as malaria, cholera and dengue. The debilitating effects of disease on the working population will hinder progress towards development in much of Sub-Saharan Africa.

Typical mistake

It is often assumed that famine means an overall shortage of food, resulting in undernutrition, malnutrition and, in the most severe circumstances, starvation. In fact food is often available during famines but is simply unaffordable to the poorest members of society with the fewest entitlements.

Typical mistake

Only rarely does desertification result in once-productive farmland becoming true desert. More often it means a significant reduction in agricultural productivity.

Now test yourself

Tested

20 Why is global warming occurring most rapidly in the Arctic?

21 Why is global warming likely to have the greatest human impact in Africa?

Answers on p. 121

The indirect impacts

Revised

Eustatic sea level rise

Global warming will result in a worldwide (or **eustatic**) rise in sea level. This is due to:

● melting of ice sheets in Antarctica and Greenland and mountain glaciers

● the thermal expansion of the oceans

In total the world's ice sheets and glaciers store enough water to raise sea level by at least 65 m. The latest estimates suggest a rise in sea level of more than 1 m by 2100. This is twice as great as the forecast made by the Intergovernmental Panel on Climate Change (IPCC) in 2007. Globally, 200 million people living within 1 m above present sea level, are at risk. For low-lying countries such as Bangladesh, small island states in the South Pacific Ocean and mega cities in the developing world (Figure 1.18), a sea level rise of just 1 m could spell disaster.

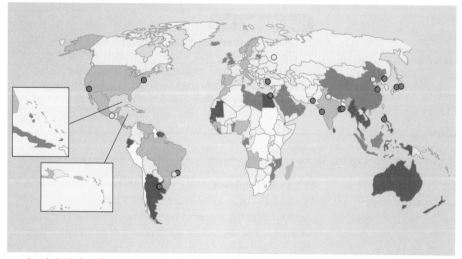

Population in low elevation coastal zones (LECZ) (%)   Mega cities

    <2   2–5   5–10   10–20   ○ Outside LECZ

    20–50   >50   Landlocked countries/No data   ● Inside LECZ

Figure 1.18 Population and mega cities at risk from a rise in sea level (Source: World Bank)

Now test yourself   Tested ☐

22 (a) What is meant by 'eustatic' sea level rise?
   (b) Give two reasons for the current eustatic rise in sea level.

Answers on p. 121

Bangladesh

The overall impact of rising sea level will be greater in Bangladesh than in any other country in the world (Figure 1.19). This is because:

- most of the country occupies the floodplains and delta of the Ganges and Brahmaputra river system, and is less than 10 m above sea level
- 15–20 million people live in areas less than 1 m above sea level; in the next 80–90 years these people will be forced to abandon their homes as rising seas permanently flood 15% of Bangladesh's total land area
- the Ganges–Brahmaputra delta is slowly sinking, so the relative change in sea level by 2100 could be more than 1 m
- rising sea level will increase the **salinity** of groundwater and soils — a particular problem in the delta where most people depend on agriculture for their livelihoods; by 2100 an estimated 60,000 ha of agricultural land could be lost to rising sea level and salinity problems
- rising sea level will destroy Bangladesh's coastal mangrove forests, which provide protection against storm surges, are a source of timber, and a vital nursery for fish and other marine life
- tropical cyclones are a major natural hazard in the coastal regions and can penetrate 100 km inland. In 1970 and 1991 they caused huge loss of life. Higher sea levels will make tropical cyclones and storm surges even more deadly
- people living on the coast and in the delta are extremely vulnerable to cyclone hazards: 44% of Bangladeshis live in poverty, 4.5 million people (mainly rural dwellers) are landless

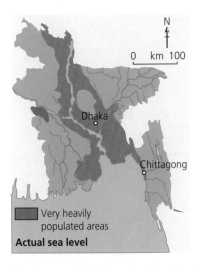
■ Very heavily populated areas
Actual sea level

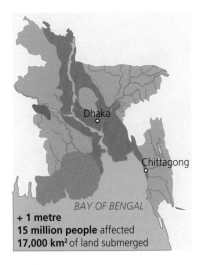
BAY OF BENGAL
+ 1 metre
15 million people affected
17,000 km² of land submerged

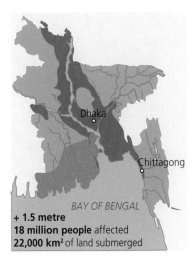
BAY OF BENGAL
+ 1.5 metre
18 million people affected
22,000 km² of land submerged

Figure 1.19 The effect of sea level rise in Bangladesh

South Pacific islands

Rising sea level is a major hazard to small island states in the South Pacific Ocean (Figure 1.20). Scattered throughout Micronesia, Melanesia and Polynesia, tiny nations such as Tuvalu and the Marshall Islands consist of coral **atolls**, barely 1 m above sea level.

Apart from their low elevation, the South Pacific islands are vulnerable to sea level rise because:

- they are exposed to tropical cyclones and storm surges
- their economies depend heavily on local natural resources such as fish and food crops

- population and economic activity are concentrated on or near the coast
- fresh water resources are scarce and easily contaminated by salt water incursion
- they have limited human and financial resources

Although the impact of sea level rise varies from island to island, there are a number of common themes.

- Predicted sea level rises of 1–2 m in the next 50–100 years will swamp some atolls. In Tuvalu, 'washovers' already occur on some islands during highest tides. Before the end of the century these islands will be uninhabitable. Evacuations have already occurred in Vanuatu, Kiribati and Samoa. In Tuvalu many islanders have emigrated to New Zealand.

- Where geography permits, coastal settlements will be forced to retreat inland. However, many islands do not have this option — there is simply no high ground to retreat to.

- Sea level rise (along with warming ocean waters and increased acidification) will damage coral reefs and increase coastal erosion. Coral ecosystems are major sources of food for many islanders.

- Food production is threatened by salt water contamination of groundwater. Lacking surface water, most islands depend entirely on groundwater for irrigation, domestic use, industry and tourism.

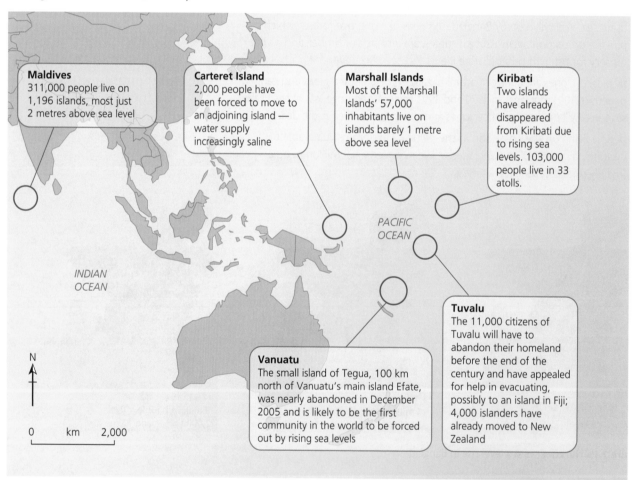

Figure 1.20 South Pacific islands threatened by rising sea levels

Maldives
311,000 people live on 1,196 islands, most just 2 metres above sea level

Carteret Island
2,000 people have been forced to move to an adjoining island — water supply increasingly saline

Marshall Islands
Most of the Marshall Islands' 57,000 inhabitants live on islands barely 1 metre above sea level

Kiribati
Two islands have already disappeared from Kiribati due to rising sea levels. 103,000 people live in 33 atolls.

Vanuatu
The small island of Tegua, 100 km north of Vanuatu's main island Efate, was nearly abandoned in December 2005 and is likely to be the first community in the world to be forced out by rising sea levels

Tuvalu
The 11,000 citizens of Tuvalu will have to abandon their homeland before the end of the century and have appealed for help in evacuating, possibly to an island in Fiji; 4,000 islanders have already moved to New Zealand

Now test yourself

Tested ☐

23 Describe the likely environmental and socio-economic impacts of rising sea level in the South Pacific region.

Answer on p. 121

Typical mistake

Only in extreme cases will rising sea level in the next 50 years result in 'washover' and the abandonment of South Pacific islands. The main effects of rising sea level will be to reduce food production, increase unemployment, contaminate water supplies, and damage coral reefs and mangroves.

Predicting the impact of climate change

Computer models

Powerful supercomputers run mathematical models that predict changes in global and regional climate in the medium and long term. Global climate models (GCMs) include those developed by the UK Meteorological Office at the Hadley Climate Research Centre. The most recent climate models combine atmospheric and oceanic systems, as well as the carbon cycle, vegetation changes and ocean biology. The Intergovernmental Panel for Climate Change (IPCC) summarises the predictions of around 20 GCMs every 6 years in its Assessment Report. Its next report is due in 2013.

Scenarios

Despite their sophistication, predictions from climate models are highly variable. For example, the latest Met Office model forecasts for average global temperatures in 2100 range from +1.6 to +4.3°C.

This uncertainty reflects:

- the complexity of the atmosphere
- the coupling of the atmosphere with other physical systems and with human systems
- **feedback** loops (e.g. melting of sea ice increasing the absorption of solar radiation and air temperature) which are poorly understood or even unknown
- the assumptions made about future population growth, energy production and consumption, and GHG emissions

Climate change at regional level is most relevant to people's lives but currently is the hardest to predict.

Examiner's tip

Be critically aware of the uncertainty of predictions of climate change. The atmosphere is complex and often behaves randomly.

Global warming and catastrophic change

Some scientists believe that increases in atmospheric CO_2 will eventually reach a critical **tipping point**, triggering abrupt and irreversible climate change.

The so-called **runaway greenhouse effect** is an example of such a scenario. Rising sea surface temperatures (SSTs) in the tropics could reach a theoretical tipping point of 30°C. Warmer oceans increase evaporation, and load the atmosphere with water vapour (a potent GHG). This change causes global temperature to rise as the planet absorbs more radiation than it re-radiates to space. Meanwhile **positive feedback** leads to a catastrophic cycle as further evaporation induces further temperature rises, etc.

Scenarios of catastrophic and irreversible change

In theory, anthropogenic global warming could cause sudden and catastrophic changes to the planet's natural systems such as ocean circulation, climate and the carbon cycle.

Collapse of ocean circulation in the North Atlantic

The North Atlantic conveyor. Ocean currents move surplus heat from the tropics towards the poles and help to maintain the global energy balance. This **thermohaline circulation** of the oceans could be disrupted by global warming.

For its latitude, northwest Europe has an exceptionally mild climate. This is due to the North Atlantic Drift, a surface ocean current that transfers warm water from the Gulf of Mexico across the Atlantic Ocean towards Iceland, Norway and the British Isles. The North Atlantic Drift is a natural conveyor delivering huge amounts of heat energy to northwest Europe. It keeps the coast of northwest Europe ice-free throughout the winter and maintains winter temperatures 10–15°C higher than the average for the latitude.

Positive feedback in a system occurs when change induces further change and instability. Negative feedback dampens change, and restores systems to stability.

Now test yourself

24 Explain with examples:
   (a) tipping points
   (b) positive feedback

Answers on p. 121

Tested

As the warm surface water drifts across the Atlantic it cools, becomes more saline and more dense. This results in **downwelling** off the coast of Iceland and Greenland. Cold, saline water sinks to the ocean floor and eventually returns south as a deep water current. Downwelling thus acts as a giant pump, powering the entire thermohaline circulation in the North Atlantic.

Observations suggest that in recent decades this downwelling has weakened and that the volume of water carried by the conveyor has dropped by 30%. There are two possible explanations, both linked to global warming:

● rapid melting of the Greenland ice sheet releasing huge volumes of fresh water into the North Atlantic

● increased discharge of major Siberian rivers such as the Yenisey, Lena and Ob, swollen by melting permafrost and higher rainfall

These large inputs of fresh water dilute the surface ocean waters, reducing their salinity and density and weakening downwelling. Eventually if a **tipping point** is reached, the pump could switch off permanently.

Impact. Changes to the thermocirculation in the North Atlantic would drastically alter the climate of northwest Europe. Without the warming influence of the North Atlantic Drift the region could plunge into deep freeze, with winters as severe as those in eastern Canada. Sea ice would form around the coastline in winter; transport systems would be disrupted; and with a shorter growing season food production would decline steeply.

Climate change in the Arctic

Causes of warming in the Arctic. The Arctic is warming faster than any other region. Computer models suggest an average rise in Arctic temperatures of at least 7°C by 2100. This rapid warming is mainly due to **ice–albedo feedback**.

● Large areas of land and sea in the Arctic are covered either permanently or seasonally by snow and ice.

● Snow and ice reflect most incident solar radiation, helping to maintain low temperatures at the surface.

● Global warming is reducing the snow and ice cover, exposing more land and water.

● Land and water have lower albedos than snow and ice, and absorb a larger proportion of incident solar radiation.

● More heat absorption results in rising temperatures, more melting, more heat absorption and so on (i.e. positive feedback).

In Siberia and northern Canada, permafrost is already melting. Evidence of melting includes thaw lakes that expand every summer, increasing discharges from the main Siberian rivers, and forest trees, telephone and overhead power masts tilting at precarious angles.

> **Typical mistake**
>
> Although tipping points induce abrupt change, this is a relative term and should not be exaggerated. Changes that occur over decades, rather than on geological timescales, are often termed 'abrupt' or 'sudden'.

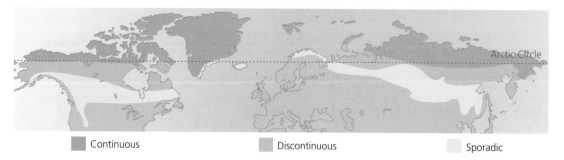

■ Continuous	■ Discontinuous	Sporadic

Figure 1.21 The global distribution of permafrost

Impact. Permafrost melting is a global as well as a regional problem. It has the potential to release huge volumes of methane (CH_4) and CO_2 to the atmosphere. Both gases, derived from plant remains frozen in the permafrost, are powerful

GHGs. Methane is of particular concern because it is 20–25 times more effective as a GHG than CO_2.

Melting of the permafrost will be irreversible. Huge increases in atmospheric CH_4 could raise global temperatures far above those predicted by current climate models, causing climate chaos and accelerating sea level rise.

Rapid warming in the Arctic will also have ecological and economic effects. These changes include vegetation and habitat change, disruption to animal migrations, the destruction of traditional indigenous economies and cultures, damage to infrastructure (e.g. buildings, roads, pipelines) and threats to land-based oil and gas extraction.

> **Examiner's tip**
>
> Questions on global warming might require knowledge of either environmental or socio-economic effects, or both.

Examiner's summary

Climate change

✔ Present-day global warming should be viewed in the context of long-term climate change over centuries and millennia.

✔ The impact and severity of climate change will show great spatial variability.

✔ Some case studies in Unit 1 are listed in the specification and are therefore prescriptive. These studies must be revised thoroughly.

✔ Some effects of global warming (e.g. desertification, drowning of small island states) are less dramatic than the

terms suggest. Exam answers must convey an appreciation of the complexity and reality of the issues and their impact.

✔ 'Tipping points' may take several decades and will appear gradually rather than suddenly.

✔ Answers to questions on the impact of climate change must consider a range of effects, including environmental, social and economic.

✔ It is important to use accurate terminology and to learn the meaning of technical terms emboldened in these revision notes.

Coping with climate change

Climate change strategies — Revised

Responses to climate change can be divided into two strategies: (1) those that tackle its root causes by reducing GHG emissions, and (2) those that seek to adapt to climate change.

> **Enquiry question**: What are the strategies for dealing with climate change?

Tackling the causes of climate change

Strategies to tackle the causes of climate change and reduce carbon emissions operate at international, national and local scales.

International strategies

Climate change — 'arguably the world's greatest hazard' — affects all countries. Solving the problem therefore requires an international effort, with individual countries setting aside national interests and cooperating for the common good.

So far, the only truly international initiative to tackle global climate change is the **Kyoto Protocol** (1997). Under Kyoto, most rich countries agreed to legally binding reductions in their CO_2 emissions (based on 1990 levels) by 2012. Kyoto exempted developing countries (including major polluters such as China and India). A number of rich countries, notably the USA, refused to ratify the treaty. Kyoto expired in 2012 and despite intensive negotiations only the EU committed to a further extension until 2015.

> **Examiner's tip**
>
> International strategies to tackle climate change should be viewed critically, e.g. the problems and obstacles to agreement.

At the UN climate conference at Durban in 2011 agreement was reached on a global Green Climate Fund. The Fund, financed by rich countries, will transfer $US100 billion to developing countries by 2020, to help them cut GHG emissions and adapt to climate change.

Cap and trade offers an alternative, international market-based approach to limit CO_2 emissions. Under this scheme, businesses are allocated an annual quota for their carbon dioxide emissions. If they emit less than their quota they receive **carbon credits**, which can be traded on international markets. Companies that exceed their quota must either purchase additional credits or pay a financial penalty.

National strategies

It is easier for individual governments to implement their own strategies to control carbon emissions. Governments may, for instance:

- give subsidies to promote cleaner technologies such as renewable, carbon-free energy (wind, solar power, nuclear power)
- impose carbon (or green) taxes on activities that consume large amounts of fossil fuel, e.g. coal-based power generation, steel production, air transport
- give financial incentives to households to invest in energy conservation and alternative energy (e.g. insulation, solar panels)

Local strategies

More and more cities are taking steps to reduce their **carbon footprints**. As 80% of global GHG emissions originate in cities, action at this scale is vital. London is tackling its emissions by:

- a congestion charge (since 2003) on road traffic entering central London, to cut emissions and reduce its impact on the environment
- a low emissions zone (2008), levying a charge on commercial vehicles in Greater London
- replacing older diesel buses with a fleet of hybrid buses (diesel-electric), with 30% lower carbon emissions
- a bicycle hire scheme to encourage environmentally friendly transport in the city centre
- recycling, and reducing carbon emissions from public buildings under the authority's control

Other local strategies that contribute to lowering carbon emissions in the UK include micro-hydro schemes, green roofs and planting trees.

Adapting to climate change

Adapting to climate change is a strategy that acknowledges the reality of climate change and the need for urgent action. The residence time of CO_2 in the atmosphere is around 100 years, which means that even if global CO_2 emissions fall within the next two decades (highly unlikely) global warming will continue until well into the next century.

Urban environments

By 2070, Chicago and New York will have summers similar to those of the southern state of Alabama. Already these cities are planning ahead, modifying urban climates by:

- replacing native trees such as White Oak and Norway Maple with drought-resistant species such as Sweet Gum and Swamp Oak
- planting more trees in streets and parks, and vegetation on rooftops, to provide shade and cooling, thus reducing the **urban heat island** effect
- installing air conditioning in city schools
- replacing tarmac and concrete pavements with permeable materials that reduce runoff and conserve soil moisture and urban water supplies

> **Now test yourself**
>
> 25 Suggest three ways in which a government can reduce greenhouse gas emissions.
>
> **Answer on p. 121**
>
> Tested

> **Typical mistake**
>
> Even if carbon emissions were stabilised within the next few years, global warming and climate change would continue for many decades. Global temperatures are certain to rise, it's just a question of by how much.

> **Urban heat island** is the phenomenon of higher temperatures in towns and cities compared with the surrounding countryside. Heat islands form at night under still-air conditions. They mainly result from the greater absorption of solar radiation by urban surfaces, heat generated by combustion and air pollution.

> **Now test yourself**
>
> Tested
>
> 26 How are cities planning to adapt to warmer climates in future?
>
> **Answer on p. 121**

Agriculture

Agriculture can adapt to higher temperatures, reduced rainfall and more extreme weather by:

- the breeding and introduction of **drought-resistant crops**
- introducing **zero tillage** (i.e. growing crops without ploughing the soil), which conserves soil moisture and increases the soil's organic content
- extending irrigation to more farmland
- diversifying into novel types of farming (e.g. viticulture, combining agricultural and forestry activities) and non-agricultural activities
- improving the effectiveness of pest and weed control

Flooding and coastal erosion

More extreme rainfall events, rising sea level and stormier conditions will increase flood risks and coastal erosion. In the UK, the Environment Agency and local authorities are already adapting to this scenario by:

- implementing **coastal management strategies** that, in some places, will result in the abandonment of costly hard sea defences (and even some settlements); this allows coastlines to retreat and form natural (and sustainable) defences such as salt marshes and mudflats
- reducing runoff through **afforestation** in upland catchments, protecting wetlands and stricter planning controls to prevent development on floodplains

Managing climate change

Revised

Table 1.5 Top ten CO_2 emitters in 2009

	CO_2 emitted (M tonnes)	Percent change 1990–2009	Tonnes/capita
China	6,832	209	5.2
USA	5,195	7	16.9
India	1,586	172	1.4
Russia	1,533	−30	10.8
Japan	1,093	3	8.6
Germany	750	−21	9.2
Iran	533	197	7.3
Canada	521	20	15.4
South Korea	516	125	10.6
UK	466	−15	7.5

Conflicting government policies

In order to succeed, the Kyoto Treaty demanded unprecedented international cooperation. In practice it was undermined by conflicting government policies.

- Although 192 countries ratified Kyoto, only 35 agreed targets to reduce GHG emissions that were legally binding. Between them, these 35 countries accounted for only one-third of global emissions.
- Developing countries such as China and India were exempt from Kyoto. They argued that restrictions on their carbon emissions would hamper their economic development.
- Developing countries argued that rich nations were responsible for global warming — that past industrialisation and high levels of consumption in the developed world were to blame. Thus rich countries had a moral duty to act.
- Although developing countries are now the major source of GHG emissions, carbon output per capita is much lower than in the developed world (Table 1.5).

- The USA and Australia took the view that as 60% of GHG emissions originated in the developing world, without curbs on these emissions Kyoto would make little difference. By 2009 China and India had become the world's first and third largest carbon polluters, respectively (Table 1.5).
- Although legally binding, countries that failed to meet their Kyoto targets (e.g. Canada, Japan), could opt out of the treaty without penalty.

Delegates from all countries met at UN climate change conferences at Cancun (2009), Copenhagen (2010) and Durban (2011) to discuss strategies post-Kyoto. But progress was limited on the fundamental issue of curbing GHG emissions.

At the climate change conference in Durban, Canada, Russia and Japan (all failed to achieve their emission targets) quit Kyoto and so avoided economic penalties. Meanwhile, developing countries continued their exemption from carbon cuts. In the light of this, the USA was not prepared to change its non-ratification stance. Only the EU countries agreed to a second round of emission cuts.

Business interests

Powerful business and commercial interests lobby governments to protect major carbon-emitting industries such as coal mining, electricity generation and steel making. In 2011, coal-fired electricity producers in the USA opposed an emissions cap (i.e. a limit on CO_2 emissions per megawatt of electricity). They gained significant political support on the grounds that the cap would 'drive up energy prices and destroy jobs'. The Canadian government abandoned Kyoto, describing it as 'a job-killing, economy destroying pact'. Ratifying the post-2012 Kyoto Treaty would have damaged its energy industries, especially the highly polluting Alberta tar sands.

Non-governmental organisations (NGOs)

Climate change will hit the world's poorest countries hardest. In the developing world, NGOs are key players in managing the effects of climate change at the local scale. It is at the grassroots level that more extreme weather will have the greatest impact. Community projects, sponsored by NGOs, can identify hazards and risks, and devise adaptive strategies to enable local people to cope with floods and droughts (e.g. water storage, water management), and related problems such as sanitation, disease and human health.

Groups

Groups such as the United Nations (UN), the Intergovernmental Panel (IPCC) on Climate Change, and the World Bank also play key roles in managing climate change.

The UN is the driving force behind the Kyoto Treaty and annual international conferences on climate change. Its Environment Programme (UNEP) promotes conservation and sustainability to improve people's lives. Climate change is one of six priority areas.

The IPCC, established by the UN, is the leading international body for assessing climate change. Free from political influence and based on research conducted by scientists worldwide, it provides an objective view of climate change and its likely environmental, social and economic impacts.

The World Bank provides financial and technical resources to support development in poorer countries. Among the World Bank's projects are support for low carbon growth (e.g. the development of renewable energy and GHG mitigation) in countries such as South Africa, Mexico and Brazil.

Individuals

Individual initiatives can make a positive contribution to reducing humankind's **carbon footprint**. Examples include:

Examiner's tip

It is important to appreciate that the atmosphere is a common resource, available to all countries. It is difficult to get agreement on usage when a resource is shared. Countries that agree to limit their GHG emissions will be worse off economically if others continue to pollute as usual.

Now test yourself

29 What do the following acronyms stand for: IPCC, NGO, UN, UNEP, GHG?

Answer on p. 122

Tested

Carbon footprint is the amount of carbon emitted by the consumption of fossil fuels by a nation, city, group or individual. It's measured in tonnes of CO_2.

Exam practice answers and quick quizzes at **www.hodderplus.co.uk/myrevisionnotes**

- energy conservation by lowering thermostat settings on home heating (and raising them on cooling) systems
- walking or cycling to work/school/shops
- purchasing locally grown food
- growing your own vegetables and fruit (i.e. eliminating packaging, transport)
- using recycling schemes
- buying hybrid or smaller, more fuel-efficient cars
- taking holidays at home
- using rail transport rather than flying
- insulating walls and roofs at home and eliminating draughts
- meeting household energy demands by generating electricity using roof-mounted solar panels and selling any surplus to power companies

The challenge of global hazards for the future

Increasing risk and uncertainty
Revised

At the global scale, the risks from natural hazards such as floods, droughts, earthquakes and tropical cyclones are increasing. Several factors account for this:

- more frequent and higher intensity hazards (e.g. floods, droughts)
- rapid world population growth exposing more people to natural hazards
- increases in the number of people living in poverty in some parts of the developing world (e.g. Sub-Saharan Africa (SSA))
- growing pressure of human populations on environmental resources (e.g. deforestation, soil erosion)

Food insecurity

Food insecurity exists when people's access to sufficient food to lead a healthy life is no longer assured. Food insecurity often increases sharply in the aftermath of natural disasters that affect the production, availability and affordability of food. The most severe food shortages result in **famines**.

> **Typical mistake**
>
> Although famines are often accompanied by rising mortality among the young and the old, starvation is rarely the cause of death. More often **malnutrition** and **undernutrition** reduce the body's resistance to infection and people die from a range of diseases.

Causes of food insecurity

At the global scale food insecurity is greatest in the world's poorest countries, particularly in Africa and south Asia (Figure 1.22). Short-lived food insecurity often follows natural disasters that disrupt farming and food distribution systems, and raise food prices. In the longer term, food insecurity is linked to global problems such as poverty, political conflict and climate change.

Poverty and food insecurity. Poverty often goes hand-in-hand with food insecurity, particularly in the developing world (compare Figure 1.22 with Figure 1.23 on page 42). Poor people have limited access to food because of their weak purchasing and bargaining power. During periods of food shortage prices soar and staple foods become unaffordable. In these conditions, poor people, with

> **Enquiry question**: How should we tackle the global challenges of increasing risk and vulnerability in a more hazardous world?

> **Typical mistake**
>
> Remember that, while the *proportion* of the global population living in poverty declined everywhere 1990–2008, the *number* of poor people living in SSA actually increased by nearly 130 million during this period.

> **Now test yourself**
>
> 30 State four reasons why the risks presented by natural hazards are increasing.
>
> **Answer on p. 122**
>
> Tested

> **Malnutrition** is a lack of adequate nutrition caused by an unbalanced diet (often shortages of protein and essential vitamins).
>
> **Undernutrition** is a condition caused by too low a food intake.

fewest exchange entitlements (e.g. savings, creditworthiness, assets) are most vulnerable.

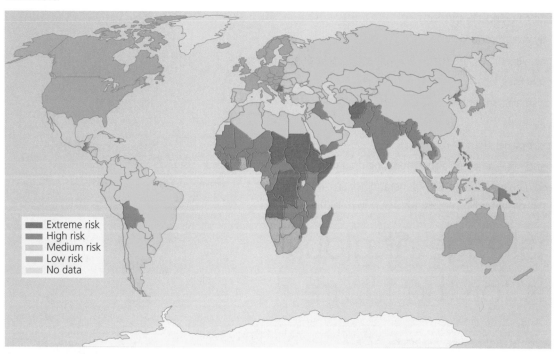

Figure 1.22 Global food insecurity (Source: Food Security Risk Index 2011, Maplecroft)

Now test yourself   Tested ▢

31 Define the following terms: food insecurity, famine, malnutrition, undernutrition.

Answer on p. 122

Examiner's tip

Food insecurity, drought and other global problems such as climate change and political conflict are often closely interconnected, e.g. food shortages, drought and political conflict in the Horn of Africa 2011.

Water shortages

Rising global demand

Fresh water is a finite resource but demand, driven by population growth and economic development, has no limits. According to the World Bank, some 1.1 billion people in the developing world have inadequate access to water. The mismatch between supply and demand is most acute in densely populated, semi-arid regions in Sub-Saharan Africa, the Middle East, India and China.

Global supply problems

Meeting future global demands for water is complicated by climate change, melting glaciers, sea level rise and political conflicts.

Climate change. Climate change in the twenty-first century is likely to reduce water supplies for millions of people. Climate models forecast that rainfall variability will increase, with extremes of flood and drought becoming more common. Also, as global temperatures rise, evaporation increases, further reducing water resources.

Any decline in rainfall or increase in its variability in semi-arid regions could make **dry farming** unsustainable. Semi-arid regions are major food growing areas. As drought takes hold in important farming regions such as the Prairies and the Pampas, cereal production could slump, causing global food shortages.

Dry farming describes farming systems that rely on direct rainfall rather than irrigation.

In south Asia, the annual **monsoon** rains are crucial for agriculture and water supply. Any weakening of the monsoon will reduce water supplies and adversely affect agriculture, food security, the environment and human health. Already there are ominous signs: the monsoon is starting later, and drier spells during the monsoon season (June–September) are increasing in length.

Melting glaciers. In Asia 1.3 billion people depend on water from glacier-fed rivers such as the Ganges and Indus. Yet 80% of Himalayan glaciers are retreating and two-thirds are expected to disappear by 2060. By the mid-twenty-first century water shortages in China and south Asia could result in food crop yields declining by up to one-third.

Sea level rise. On low-lying coasts, rising sea levels will cause salt water intrusion to pollute aquifers and the risks of flooding will increase. In the coastal areas of southwest Bangladesh 60% of arable land is already affected by salinisation. Salt water contamination of groundwater resources is also caused by storm surges driven by tropical cyclones, which are likely to increase in frequency and intensity in future.

Political conflict. Declining river flow due to climate change, together with rising demand, threaten to result in political (and armed) conflict in international drainage basins. The most common dispute occurs when an upstream state decides unilaterally to exploit additional water resources, thus reducing either the volume or quality of water to downstream users.

Typical mistake

Remember that water availability is not just determined by mean annual rainfall. Two other factors that influence water availability are the proportion of rainfall lost to evapotranspiration and rates of runoff.

Examiner's tip

When reviewing the impact of global climate change it is important to cover a broad range of issues. You should consider the far-reaching effects of climate change on economic activity, the environment, demography and international politics, etc.

Meeting the challenge of global warming

Revised

Reducing carbon emissions

Combustion of fossil fuels is the main driver of global warming and climate change. While the potential catastrophic effects of global warming have been recognised for several decades, CO_2 and other GHGs emissions have continued to rise inexorably. The UN target to reduce GHG emissions and limit global warming to less than 2°C by 2100 now has little chance of success. Failure by governments to tackle global GHG emissions is explained by:

- planning only for the short-term
- placing economic goals above environmental responsibilities
- seeing little point in making economic sacrifices when other countries continue to pollute
- opting for political expediency and either ignoring or denying the problem

Examiner's tip

It is important to clarify your own views on the best way to tackle global warming and evaluate alternative strategies. Your views should be (a) balanced, (b) based on a detailed understanding of the issue, and (c) be supported by factual evidence.

Renewable energy

Renewable energy is either **carbon neutral** or **carbon free** and sustainable. Currently around 13% of the world's energy production is from renewables, mostly from wood used for cooking and heating in the developing world. Only 2.8% of global energy comes from advanced technology renewables such as wind, solar, geothermal and HEP.

Hydro-electric power (HEP)

HEP is the most popular advanced-technology type of renewable energy. It has the potential to generate enormous amounts of electricity (e.g. China's Three Gorges 18.2 GW, Brazil and Paraguay's Itaipu 12.6 GW) but it also has disadvantages.

- Most schemes require the construction of dams, flooding large areas upstream and destroying habitats, wildlife, farmland and even settlements. Downstream the character and behaviour of rivers are changed (e.g. colder, clearer water, absence of floods), damaging habitats and ecosystems.
- The huge capital costs of large dams (the Three Gorges dam cost $US22.5 billion) and the disbenefits to local people whose livelihoods and homes may disappear.

Carbon neutral refers to energy sources where rates of CO_2 (and other GHG) emissions are balanced by rates of carbon storage, e.g. biofuels.

Carbon free energy involves no direct release of CO_2 and GHGs, e.g. solar power.

Wind power

The UK wants to generate 15% of all its electricity from renewables by 2020. To meet this target it will rely mainly on wind power. Because of the difficulty in getting planning permission for wind farms on the mainland, new wind farms are increasingly sited offshore (e.g. in the Thames estuary, the Wash, the North Sea).

Proposals for wind farms in the UK arouse strong opposition from local communities and conservationists because they:

- require large tracts of land — 7 ha for every MW of electricity
- are often sited on exposed hillsides and ridges where wind turbines are visually intrusive, and on coasts and uplands that have high amenity value
- disturb wildlife habitats and increase bird mortality

Solar power

Ideal conditions for solar energy are found in tropical and sub-tropical deserts where solar radiation is intense and skies are cloud-free. Because deserts are sparsely populated, solar power stations have few potential objectors. The development of solar power is still in its infancy, though its potential is huge.

Geothermal energy

Geothermal power stations, tapping heat from the Earth's interior, operate in Iceland, New Zealand and California. Like solar power, the potential for geothermal is huge and the energy supply inexhaustible. There are few objections to geothermal energy on environmental grounds.

Nuclear energy

Opinion remains divided on the issue of nuclear power. Nuclear energy has, however, become an increasingly attractive option because it is carbon free.

Nuclear power has several disadvantages:

- major accidents involving nuclear power (e.g. Chernobyl 1986, Fukushima 2011) have the potential to kill and injure thousands of people and cause long-term damage to the environment
- no secure long-term storage facility for radioactive waste and spent uranium fuel has yet been built by any country
- nuclear power stations and reprocessing plants are potential targets for terrorists
- high construction and decommissioning costs mean that electricity from nuclear power is more expensive than electricity generated by fossil fuels
- nuclear power plants require complex technology and may take a decade or more to build

Energy efficiency

Improving energy efficiency reduces consumption and GHG emissions. Governments legislate to reduce emissions from motor vehicles, factories, commercial activities and households (e.g. smaller cars with more fuel-efficient engines pay lower road taxes, government grants are available for cavity wall and loft insulation). Governments also encourage the use of non-petroleum fuels such as biodiesel and liquid natural gas (LNG) and promote domestic solar energy generation and micro-HEP stations. In California a carbon tax is levied on electricity generated in coal-fired power stations to encourage usage of cleaner fuels such as gas and renewables.

Reafforestation

Deforestation transfers carbon, stored in forest trees, to the atmosphere. Today deforestation accounts for nearly 20% of all GHG emissions. Two hundred years ago the **primary rainforest** covered 14% of the Earth's land surface. By 2010 the proportion had shrunk to just 6%.

Typical mistake

Remember that solar energy can be highly effective outside the tropics and sub-tropics. Germany currently generates around 4% of its electricity from solar power, and this figure will rise significantly in future.

Typical mistake

The examples of HEP and wind power show that it is wrong to think that all renewables are environmentally friendly. Moreover, although renewables are largely carbon-free, carbon emissions occur during construction and assembly.

Now test yourself

32 What is the difference between carbon free and carbon neutral energy?

Answer on p. 122

Tested

Examiner's tip

It is sensible to argue that no single option will solve the problems of global warming and climate change. Successful strategies will combine initiatives such as international agreements on GHG emissions, improvements in energy efficiency, the expansion of renewable energy, nuclear power, reafforestation, etc.

Now test yourself

33 What is the main environmental advantage of nuclear energy?

Answer on p. 122

Tested

Protecting tropical forests from loggers, farmers and miners is an inexpensive way of curbing GHG emissions. The UN's *Reducing Emissions from Deforestation and Forest Degradation* (REDD) scheme incentivises developing countries to conserve their rainforests by placing a monetary value on forest conservation.

$US30 billion of aid per year is channelled to developing countries as compensation for protecting their forests. REDD also promotes low-carbon paths to development, conserves biodiversity and secures vital **ecosystem services**.

> **Ecosystem services** are resources and processes provided 'free' by ecosystems. They include food, clean drinking water, the decomposition of organic wastes, pollination, carbon storage, etc.

Technological fixes

So far technological fixes to the problem of global warming have made limited progress. Among the proposed fixes are:

- reducing insolation by installing huge mirrors in outer space to reflect the Sun's rays
- seeding the atmosphere with droplets of sulphuric acid to increase cloudiness and solar reflection
- seeding the oceans with tiny iron particles to stimulate the growth of **phytoplankton** and algae that absorb CO_2 from the atmosphere

Most ideas are either impractical or have unknown and potentially damaging environmental side effects.

> **Typical mistake**
>
> There is a commonly held view that **nuclear fusion** will eventually meet all future energy needs. However, it is unlikely that the technology to generate electricity from nuclear fusion will be available before the mid twenty-first century at the earliest.

> **Nuclear fusion** is the process that powers the Sun. Unlike conventional nuclear power, which is based on splitting atoms (fission), nuclear fusion of hydrogen atoms produces no long-lasting radioactive waste.

One innovative technology that appears to be feasible is **carbon capture**. Carbon capture involves filtering CO_2 from industrial smokestacks and power stations, and storing it permanently deep underground in old oil and gas reservoirs. However, the technology is untried and the commercial costs are high.

Risk, vulnerability and solutions ———————————— Revised ☐

Reducing disaster risk

Vulnerability to natural hazards in developing countries is increased by heavy dependence on farming and other activities that rely on the natural environment. Disasters and poverty are linked in two ways:

- disasters, such as floods and droughts, help to cause poverty and hold back economic development
- natural disasters increase poverty (e.g. destruction of livelihoods and infrastructure, spread of disease)

Thus strategies to lower disaster risk often focus on reducing poverty.

> **Examiner's tip**
>
> The impact of natural disasters is highly variable. Natural disasters affect the poor, children, the elderly and women disproportionately. Marginal groups, with fewest assets and often relying most heavily on environmental resources (e.g. farming, fishing), are especially at risk.

Flood hazards at the local scale

Past solutions to floods and other natural hazards have often relied on a structural approach (i.e. hard engineering such as building levées and storm shelters). Today, responses are more **holistic**. They focus on wide-ranging strategies to minimise risk by reducing poverty and making communities more resilient.

In 1996 and 2001 flood disasters in Vietnam killed more than 1,600 people, damaged millions of homes and devastated 400,000 ha of farmland. In response, two NGOs, *World Vision* and the *Red Cross*, sponsored disaster-reduction programmes in Vietnam's Quang Ngai province. Completed in 2009, the programmes reduced peoples' flood vulnerability. The strategy for Quang Ngai involved:

- devising household and village disaster-reduction plans to help communities deal with flood disasters
- diversifying household incomes (e.g. fish farming, animal husbandry and small businesses) to reduce the impact of floods
- housing improvements to increase flood resistance, e.g. raised foundations, strengthened roof beams and walls
- basic infrastructure improvements such as concrete bridges, medical stations, electricity generators
- improved early warning systems

Addressing the wider problem of poverty in Quang Ngai has proved difficult. Poverty is related to the small size of many farm holdings and infertile soils. Economic diversification is also hindered by the isolation of the region from Vietnam's main markets.

> **Examiner's tip**
>
> In the relationship between poverty and natural disasters, causal connections operate in both directions. Poverty makes people more susceptible to natural disasters, while at the same time natural disasters contribute to poverty.

Regional poverty: Sub-Saharan Africa (SSA)

Poverty is overwhelmingly concentrated in the economically developing world (Figure 1.23), and especially in SSA. In 2010, 29 of the world's poorest countries were in SSA.

The UN defines poverty as 'a reduced (or complete lack of) access to materials, economic, social, political or cultural resources needed to satisfy basic needs'. The World Bank quantifies extreme poverty as living on less than US$1.25/day. In 2010 more than half the population of SSA lived in extreme poverty.

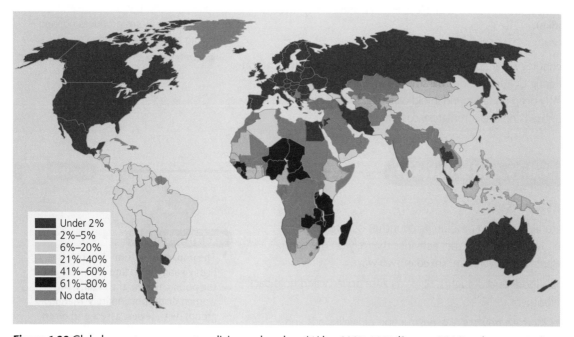

Legend:
- Under 2%
- 2%–5%
- 6%–20%
- 21%–40%
- 41%–60%
- 61%–80%
- No data

Figure 1.23 Global poverty — percentage living on less than $1/day, 2007–2008 (Source: UN Development Indices 2008)

Poverty in SSA is due to many factors such as environmental degradation, the impact of colonialism, political instability, disease, debt and so on. However, **multilateral agencies** such as the UN and World Bank are acutely aware that natural disasters are also an important cause of poverty.

> **Multilateral agencies** are international institutions such as the UN and World Bank that focus on economic and social development and the channelling of aid to recipient (developing) countries.

Between 1970 and 2010 more than 1,000 natural disasters occurred in SSA, affecting around 300 million people. Droughts and floods accounted for 80% of the loss of life and 70% of economic losses. The frequency of natural disasters in SSA reflects the vulnerability of the region's population; a vulnerability rooted in chronic poverty, land degradation, rapid urbanisation, dependence on rain-fed agriculture, political instability and weak environmental controls.

International strategies to tackle risk and vulnerability

Reducing poverty makes people less vulnerable to natural disasters and mitigates their impact. In recent years the support of multilateral agencies, governments in MEDCs, and NGOs has helped reduce the global incidence of extreme poverty. The proportion of the world population surviving on less than US$1.25/day will have halved in the period 1990–2015 to less than 15%. In SSA, the world's poorest region, poverty levels will decline to just over one-third of the total population by 2015.

Millennium Development Goals (MDGs): 2000–2015. The UN's eight MDGs set targets for reducing global poverty, hunger and gender equality, and improving standards of education, environment and health by 2015. As a spin-off, achieving (or partly achieving) these targets will go some way to reducing disaster risk in the world's poorest countries.

Examiner's tip

Natural disasters are just one of many causes of poverty in the developing world. But there is no doubt that reducing levels of poverty through social and economic development will help to lessen the impact of natural disasters.

Now test yourself

Tested ☐

34 Draw a flow diagram to show the connections between poverty, vulnerability and disaster risk.

Answer on p. 122

Examiner's summary

Coping with climate change

✔ Be able to explain why international efforts to control GHG emissions have had only limited success.

✔ Global warming and climate change is unstoppable in the next 100 years. Actions today will only influence the rate and degree of change in the long term.

✔ Most issues in geography have multiple strands and explanations. Discussions centred on a single factor (e.g. water shortages are due to low rainfall) are usually simplistic and must be avoided.

The challenge of global hazards for the future

✔ A combination of initiatives — GHG emissions, renewable energy, energy conservation — is needed to tackle the problems of global warming and climate change successfully.

✔ The human impact of natural disasters is highly variable and affects some groups more than others.

✔ Poverty and natural disasters act as both cause and effect. They are linked in a two-way, positive feedback relationship.

✔ Tackling poverty is an effective way to mitigate the impact of natural disasters in the developing world.

Exam practice

Section A

1 Study Figure 1.9 on p 18.

(a) Which region experiences the highest earthquake hazard risk? Put a tick against the correct answer. [1]

North Africa

Central Australia

Southwest Asia

Siberia

Eastern North America

(b) Describe the global distribution of earthquake hazards. [3]

(c) Give reasons for the distribution of earthquake hazards. [3]

(d) Explain the contrasting economic and social impact of earthquake hazards in rich and poor countries. [4]

2 Study Figure 1.17 on p 27.

(a) Describe:

(i) Trends in the extent of Arctic sea ice 1978–2011. [1]

(ii) Fluctuations around this trend. [1]

(b) What was the maximum and minimum extent (in km²) of Arctic sea ice between 1978 and 2011? [1]

(c) Describe how natural processes have driven climate change in the past. [3]

(d) Explain the likely environmental impact of global warming in the Arctic. [4]

Section B

3 Study Figure 1.24.

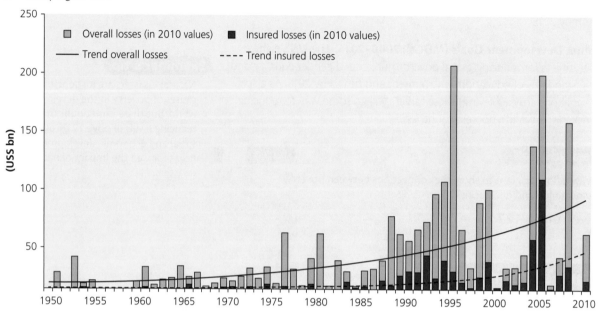

Figure 1.24 Great natural catastrophes worldwide 1950–2010 (Source: Munich Reinsurance Company)

(a) Suggest possible reasons for the global rise in natural disasters between 1950 and 2010. [10]

(b) Examine the economic and social vulnerability of California and the Philippines to natural disasters. [15]

4 Study Figure 1.25.

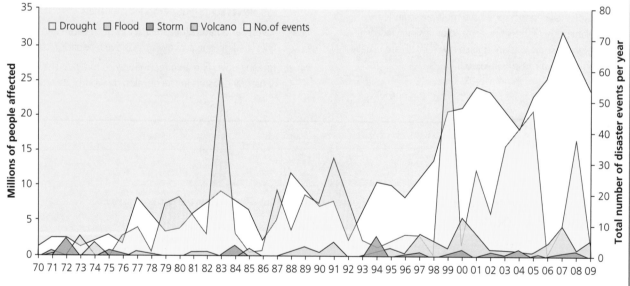

Figure 1.25 Disasters in Sub-Saharan Africa

(a) Suggest why the number of drought disasters in Sub-Saharan Africa increased between 1970 and 2009. [10]

(b) Examine the contribution of human factors to drought disasters in Sub-Saharan Africa between 1970 and 2009. [15]

Answers and quick quiz 1 online

Online

2 Going global

Globalisation

The concept of globalisation

Revised

Globalisation refers to economic and social processes that increase interdependence between countries, i.e. the flows of ideas, people, products, services and capital. In the past 50 years globalisation has brought greater integration and interdependence to the world economy, forged closer economic ties between countries, and strengthened connections between people and environments across the globe.

The following economic and social trends are closely linked to globalisation:

- the growing proportion of manufactured goods produced by **transnational corporations** (TNCs)
- large increases in international trade
- the rising importance of LEDCs to the global economy, with increased flows of raw materials, manufactured goods and components from the developing to the developed world
- the decline, due to international competition, of many traditional manufacturing industries in MEDCs, i.e. **deindustrialisation**
- the offshoring of services such as ICT, billing and customer care from MEDCs to LEDCs
- the greater dependence of the global economy on flows of capital, controlled by major **world cities**, especially New York, London and Tokyo
- rapid increases in the international **migration** of people and the creation of an international labour market

Globalisation also has a cultural dimension. Companies such as McDonald's and Starbucks market their products on a global scale, transmitting American consumer values and lifestyles.

> **Enquiry question**: What is globalisation and how is it changing people's lives?

> **Examiner's tip**
>
> Globalisation is both a process that drives economic and social change as well as a summary description of these changes.

Factors that have accelerated globalisation

Revised

TNCs in the global economy

Transnational corporations (TNCs) are large business enterprises such as Ford Motors, Nike, Apple and IKEA, with production and/or service operations in several countries and which compete in global markets. They are a major driving force behind the globalisation of industry, services and trade. Because of their economic importance, TNCs also have great political influence.

Explaining the growth of TNCs

The emergence of TNCs as dominant players in the global economy is due to three factors: their flexibility, their size and political influence, and the growth of global telecoms.

Flexibility

TNCs exploit geographical differences in the costs of production and government policies towards **foreign direct investment** (FDI). Most governments want to attract FDI and therefore offer incentives to TNCs such as reduced taxes and subsidies. However, the relationship between states and

> **Typical mistake**
>
> TNCs are not confined to the manufacturing sector. Many well-known companies in the service sector, such as McDonald's, Starbucks, Goldman Sachs and HBSC, are TNCs.

major TNCs is not always equal. Given their flexibility, TNCs can easily switch production elsewhere to lower-cost locations. For example, in 2006 Peugeot, the French TNC, located production of its new 207 model in the Czech Republic, rather than at its Coventry plant. Peugeot's motive was economic: labour costs in the Czech Republic were only one-fifth of those in the UK. The result: closure of the Coventry factory with the loss of 2,300 jobs.

Size and political influence

TNCs serve international markets. This favours large-scale production of goods and services, enabling them to achieve **economies of scale** and lower unit costs. Because of their economic power (TNCs provide employment, exports and inject spending into local economies), most governments are eager to attract investment by TNCs. As a result, TNCs are in a strong position to negotiate favourable terms and conditions, such as financial subsidies and tax concessions. The advantage of overseas locations for TNCs are: (a) lower production costs and (b) access to foreign markets that otherwise might be restricted by tariff barriers and trade quotas.

Global telecoms

Advances in telecoms have boosted globalisation and the expansion of TNCs. In the past decade, the success of global retailers such as Amazon has been based entirely on the internet and e-marketing. Global telecoms networks allow TNCs to access and share information across their geographically dispersed operations, coordinate and implement management strategies, and organise production and sales in diverse cultural and political environments.

Global markets

The huge expansion of global demand for goods and services has been a key factor in contemporary globalisation. Demand has been driven by:

- rapid population growth in developing countries (from 4.14 billion in 1990 to 5.76 billion in 2012)
- industrialisation of several large economies (**BRICS**) in the developing world, which has raised living standards among domestic consumers and increased the demand for consumer goods and services
- the emergence of market-based economies in countries such as China and Russia
- the liberalisation of trade and capital markets, with the removal of many obstacles to international trade such as import tariffs, import quotas and subsidies
- falling transport and communication costs (e.g. huge bulk carriers and container vessels) enabling firms to locate different stages of production in different countries
- telecoms and the growth of the internet and e-commerce increasing information flows between producers and consumers

International organisations

A number of international organisations such as the International Monetary Fund (IMF), the World Bank (WB), and the World Trade Organisation (WTO) play key roles in globalisation (Table 2.1). Their involvement supports the notion that globalisation is a positive process, bringing more benefits than disbenefits to the world's poorest countries.

Table 2.1 International organisations and globalisation

IMF	Assists the world economy by encouraging countries to adopt sound economic policies, e.g. exchange rates and currency values. In this way it promotes world trade. The IMF also gives advice and makes loans to member countries.
WB	The WB's purpose is to combat poverty by promoting economic development. It provides loans, advice and technical help for development projects in poor countries.
WTO	The WTO promotes international free trade and opposes protectionism. It enforces free trade rules and helps countries fight barriers to trade.

> **Now test yourself**
>
> 1 What is foreign direct investment?
>
> **Answer on p. 122**
>
> Tested

> **Economies of scale** are savings in unit costs that arise from large-scale production. They derive from fixed costs (e.g. local taxes, research and development) being spread over a larger number of units produced, greater specialisation of labour, and discounts for purchasing materials and/or services in bulk.

> **Typical mistake**
>
> Globalisation is not a new process. Ford established its first overseas plant in the UK in 1911. Contemporary globalisation began after the Second World War, and accelerated from the mid-1980s.

> The so-called **BRICS** economies are Brazil, Russia, India, China and South Africa.

> **Examiner's tip**
>
> Globalisation is not due to a single factor. Convincing explanations will refer to a range of factors such as TNCs, transport and communications, international organisations, the liberalisation of world trade, etc.

Honda, founded in 1948, is a large Japanese TNC. Its core activity is automobile manufacture. Today, Honda is Japan's second largest automobile maker (after Toyota) and the world's fifth largest. In 2010 its automobile sector alone had a turnover worth US$106 billion — more than the total GDP of Bangladesh. Like many other large TNCs, Honda is a diversified company. In addition to its automobile sector, it is the world's biggest manufacturer of motorcycles and automotive engines. It also makes robots, garden equipment and provides financial services. Worldwide, Honda employs 180,000 people.

Now test yourself

2 What do the following acronyms stand for: TNC, BRICS, IMF, WB, WTO?

Answer on p. 122

Tested

Globalisation and population movements

Revised

Globalisation has been accompanied by unprecedented population movements both within and between countries. In 2011 the world had 215 million first-generation international migrants — 40% more than in 1990 — or 3% of the world's population.

Globalisation contributed to spectacular increases in **migration** because:

- economic development, especially in emerging economies, has fuelled the demand for labour in towns and cities, resulting in massive rural–urban migration

- economic growth and rising prosperity in MEDCs in the past 50 years have created shortages of unskilled labour which have been met by foreign immigrant workers; meanwhile shortages of highly skilled graduates have encouraged flows of international migrants (from both MEDCs and LEDCs) employed in sectors such as healthcare and financial services

- booming economic growth in China, south and east Asia and the Middle East has generated huge numbers of international students who become temporary migrants in the USA, Canada, the UK and other developed countries. By 2025 the annual global number of international students (now 4 million) is set to double

- closer integration of national economies with the formation of supra-national bodies such as the EU have allowed the free movement of people between member states

- improvements in transport (falling travel costs) and modern telecoms (e.g. e-mail, skype, etc., allowing people to stay in touch with home), encourage migration and facilitate flows of information to potential migrants

Internal migration in China

In the past 30 years, millions of Chinese have migrated from the rural interior to the coast. *The Economist* magazine called it 'the greatest wave of voluntary migration in history'. Migrants have provided the labour for China's export-oriented industries, making the country the twenty-first century's 'workshop of the world'. Hotspots of immigration included the coastal provinces of Guangdong, Zhejiang and Shanghai (Figure 2.1). Because of migration Shanghai's population almost trebled between 2000 and 2010 and around 60% of Shanghai's 7.5 million young adults are migrants.

One effect of internal migration in China has been explosive urbanisation. Between 2000 and 2010 China's urban population increased by 205 million. China now has the world's largest urban population. During the same decade, the percentage of urban dwellers grew from 37% to nearly 50%.

Rising land and labour costs in southeast China and the global recession (weakening demand for Chinese exports) will reduce the scale of internal migration in future. At the same time, investment in interior provinces such as Sichuan will lessen the concentration of manufacturing on the coast. In 2008, 58% of Sichuan's 20 million migrants worked outside the province. By 2011 this figure had fallen to 52%.

> **Migration** is a permanent or semi-permanent change of residence. At the large scale it may be intranational or international.

> **Typical mistake**
>
> International migrants, moving for economic reasons, are not just unskilled workers from the developing world. The globalisation of industries and services has created an international labour market for highly qualified workers moving between developed countries or developed countries and emerging economies.

Now test yourself

3 What is meant by the term *international migration*?

4 Describe three features of internal migration in China in the past 20 years.

Answers on p. 122

Tested

> **Examiner's tip**
>
> When referring to case studies don't forget to quote specific places, patterns, trends and statistics.

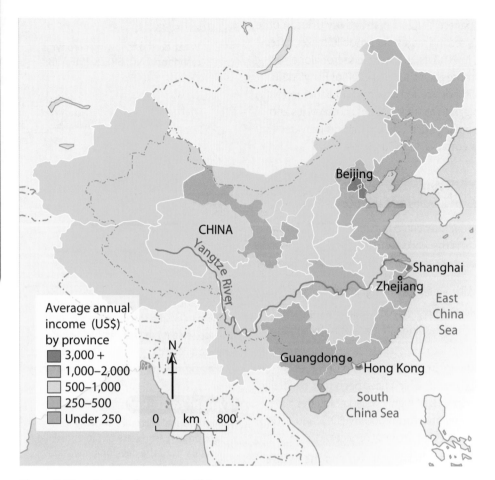

Figure 2.1 Immigration hotspots in China

Immigration to the UK since 2004

Between the late 1990s and 2012 the UK experienced an unprecedented wave of immigration (Figure 2.2). During this period annual **net migration** averaged around 250,000 people, with approximately 60% of immigrants coming from outside the EU. Today approximately one in every eight UK residents was born outside the country.

> **Net migration** is the difference between immigration and emigration over a defined period.

Large-scale population movements both into and out of the UK are linked to globalisation. EU citizens have a legal right to live and work in member states and immigration to the UK rose sharply following the accession of the EU8 countries in 2004. Factors contributing to the flood of EU8 immigrants into the UK were:

- the absence of immigration controls — temporarily waived by the UK government
- the demand for labour in the UK's booming economy
- the wide economic disparities between the UK and eastern Europe, e.g. GNI per capita in the UK was four times greater than in Poland, unemployment in Poland was twice as high as in the UK, and the minimum wage in the UK was more than five times greater
- more generous social benefit entitlements in the UK compared with EU8 countries

But not all immigrants to the UK are economic migrants. The 430,000 foreign students in higher education in the UK in 2011 (of whom 70% were from non-EU countries) accounted for a significant proportion of international

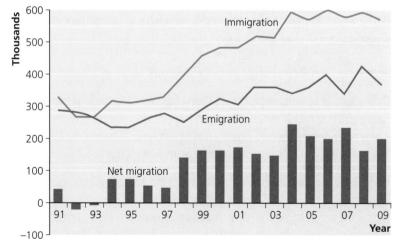

Figure 2.2 UK international migration 1991–2010 (Source: ONS)

immigrants. Higher education is a lucrative, globalised industry, where the UK competes with other developed countries for overseas students.

Despite big increases in EU immigrants, the New Commonwealth countries, led by India and Pakistan, remain the biggest single source of immigrants. In 2011 an estimated 170,000 New Commonwealth immigrants entered the UK, two-thirds of them to study.

Political refugees seeking asylum in the UK to escape war and persecution are an important immigrant group and another indicator of an increasingly interdependent and globalised world. Currently asylum seekers average about 25,000 a year. The main flows of asylum seekers in 2011 were from political trouble spots in Afghanistan, the Sudan, Somalia, Nigeria, Pakistan and Sri Lanka.

Examiner's tip

International migration is a two-way process. While nearly 600,000 people migrated into the UK in 2010, 350,000 emigrated. Bear in mind also that many immigrants (e.g. foreign students) only settle on a semi-permanent basis and eventually return home.

Now test yourself

5 How and why did immigration to the UK change after 2004?

Answer on p. 122

Tested

Global groupings

Economic and political groupings   Revised

Countries are classified into a number of broad economic and political groupings that often reflect disparities in wealth and poverty. These groupings may be informal generalisations such as 'North' and 'South', or formal politico-economic organisations such as the European Union (EU) and the North American Free Trade Agreement (NAFTA).

Enquiry question: What are the main groupings of nations and what differences in levels of power and wealth exist?

North and South

At a global scale, the terms rich 'North' and poor 'South' provide a convenient summary of global economic development. The so-called **north–south divide** is delimited geographically by the **Brandt Line**. However, this dichotomy is only a broad generalisation. Many of the world's poorest countries, especially in Africa and Central America, are in the Northern Hemisphere, while rich countries such as Australia and New Zealand, and middle income countries such as Chile and Argentina, are in the South.

Now test yourself

6 Why is the term 'North–South' not a particularly accurate description of the geography of global development?

Answer on p. 122

Tested

More, less and least economically developed countries

Level of economic development is most widely used as the criterion for grouping countries. Three groups are recognised:

- more economically developed (MEDCs)
- less economically developed (LEDCs)
- least developed (LDCs)

MEDCs

MEDCs have relatively high **per capita GDPs** and incomes. Poverty is usually confined to a minority and most citizens enjoy a high standard of living. Most MEDCs are in Europe, North America and east Asia.

LEDCs

LEDCs have much lower per capita GDPs and incomes. A large proportion of their populations live in poverty and have limited access to jobs, affordable housing and basic services.

LDCs

Within the LEDC grouping the poorest countries are identified as least developed or LDCs. In 2010, 43 out of 213 countries were classed (by the UN) as 'low income' and 28 were in Sub-Saharan Africa (SSA). These countries are characterised by:

Per capita GDP is the most widely used measure of economic development. It is calculated by dividing the total value of goods and services produced in a country or region in a year by its population.

- poor nutrition, poor health, and inadequate education and adult literacy levels
- economic vulnerability, with dependence on primary production, trade instability and indebtedness
- vulnerability to natural hazards, poor governance and political instability

The World Bank (WB) uses a more objective classification of development based on per capita gross national income. This system defines four development categories: high income (>$US12,275 per year), upper middle income ($US3,976–12,275 per year), lower middle income ($US1,006–3,975 per year) and low income (<$US1,006 per year). In response to economic change, the WB classification is updated each year. For instance, in 2011, China's rapid economic development resulted in its upgrading from 'lower middle income' to 'upper middle income'.

Although a convenient shorthand, the WB's classification is highly generalised. It hides the fact that economic development is a continuum. Rich and poor countries only occupy the extremes of the distribution and most countries are positioned somewhere in the middle. It also ignores social development. Cuba, for example, with a relatively low per capita gross income, provides high quality and free education and healthcare to all its citizens.

Newly industrialising countries and emerging economies

Countries belonging to the same WB grouping have not necessarily reached the same stage of development. Several countries, currently undergoing rapid economic development, are labelled **newly industrialising countries** (NICs) or **emerging economies**. The best known examples are the so-called BRICS economies of Brazil, Russia, India, China and South Africa.

Trade organisations

Many countries combine to form regional **trade blocs** such as the European Union (EU) and the North American Free Trade Agreement (NAFTA). One-third of the world's trade takes place within regional trade blocs. Trade blocs promote the interests of member states by:

- encouraging free trade between their members
- protecting member states' industries and services from foreign competition by using tariffs, quotas and subsidies

The EU

The EU is the most economically integrated of the world's trade blocs. It promotes the interests of its members by:

- removing internal barriers to trade (e.g. tariffs), and the movement of capital and people between its members
- a common currency (adopted by 17 members), which eases the movement of goods, services and people across internal borders
- protecting its own industries from foreign imports by imposing a **common external tariff** and **quotas**
- subsidising the exports of selected industries

Agriculture is the most heavily protected industry in the EU. In addition to tariffs on foreign food imports some agricultural exports are made artificially competitive in foreign markets by **subsidies**.

OPEC

The Organisation of Petroleum Exporting Countries (OPEC) is an association of 12 of the world's largest oil-exporting countries. OPEC regulates oil supply to meet global demand. OPEC has often been accused of operating a cartel because it maintains oil prices at levels in the interests of its members (sometimes higher than demand would justify).

> **Examiner's tip**
>
> While the classification of countries into MEDCs, LEDCs and LDCs is a useful shorthand, answers to essay-type questions must convey the idea that economic development is more complicated than this simple three-way division implies.

> **Now test yourself**
>
> 7 Which is the world's least developed region?
>
> **Answer on p. 122**
>
> Tested ☐

> **Typical mistake**
>
> The EU is more than an economic or trading union. Ever-closer economic integration requires stronger political ties. This was illustrated by the debt crisis in the Eurozone in 2012 that showed the need for a common fiscal policy to support the common currency.

OECD

The Organisation for Economic Cooperation and Development (OECD) is an international forum for discussion of economic policy issues and environmental, agricultural and energy concerns. It comprises 34 member countries, including most of the world's advanced economies and some emerging economies such as Mexico and Chile. All OECD member states are committed to democracy and free trade. The OECD's policies aim to improve the economic and social well-being of people around the world, and promote sustainable economic growth.

G8 and G20

The group of 8 (G8) nations, founded in 1975, is a forum for the governments of the world's largest economies (excluding China and Brazil). Its current composition is France, Italy, Germany, UK, Canada, USA, Japan and Russia. Since 2009 the G8 has been effectively replaced by the G20 as the main economic council of wealthy nations.

Now test yourself

8 What do the following acronyms stand for: LEDC, MEDC, LDC, NAFTA, NIC, OPEC, OECD?

9 How does free trade boost globalisation?

Answers on p. 122

Tested

The role of TNCs
Revised

The dominance of TNCs in the global economy is summarised as follows:

- they are responsible for four-fifths of global economic output and 30% of world GDP
- the top 500 TNCs account for 90% of **foreign direct investment (FDI)**
- they generate two-thirds of world trade, with one-third of world trade comprising intra-firm trade between TNCs
- they are important employers: globally US TNCs employed 24 million workers in 2010
- 53 of the world's largest 100 economic organisations were TNCs (in 2012) — the other 47 were countries

Because TNCs control flows of capital, materials, components and technical expertise — all of which are crucial to economic development — individual governments are reluctant to challenge them. Although inward investment by TNCs has considerable economic and social advantages, it also creates disadvantages (Table 2.2).

Table 2.2 Social and economic impacts of inward investment by TNCs on national economies

Advantages	Disadvantages
Creates employment for local people.	Many jobs are low-skilled in labour-intensive industries (electronics, clothing). In MEDCs capital-intensive, foreign enterprises may offer few higher paid jobs, e.g. managerial, development, design and marketing.
Rising incomes increase living standards among employees.	Lack of security, as TNCs switch operations to lower-cost locations elsewhere. Decisions driven by company budgets, rather than the social and economic interests of local communities.
Boosts exports and helps the trade balance.	Lack of governmental control, with key investment decisions taken overseas at company headquarters.
Develops and improves skill levels and expertise among the workforce as well as technology and process systems among local firms.	TNCs may demand further government financial incentives not to disinvest.
Increases spending and creates a multiplier effect within local economies.	Competition can lead to the closure of domestic firms.
Attracts related investment by suppliers to create clusters of economic activity.	

Examiner's tip

Remember that large TNCs are controversial, bringing economic and social benefits and disbenefits to communities. You should be aware of the pros and cons of inward investment by TNCs and be prepared to argue a point of view, supported by evidence and examples.

Now test yourself

10 State four advantages of foreign inward investment by TNCs.

Answer on p. 122

Tested

Case study   **Closure of Hoover-Candy's factory at Merthyr Tydfil, 2009**

Background. Hoover-Candy is an Italian TNC with headquarters near Milan. The company employs 7,610 people worldwide, 6,000 outside Italy. It has 38 branch plants. It manufactured washing machines and tumble dryers at its Merthyr Tydfil plant. The plant was originally opened by the US TNC, Hoover, in 1948. The Merthyr Tydfil plant, an early example of FDI, employed more than 5,000 workers at its peak.

Closure. Hoover-Candy closed its Merthyr Tydfil plant in March 2009. Closure was justified on grounds of high operating costs, making it uncompetitive with plants in China and eastern Europe. Production was transferred to a lower-cost location — Turkey.

Impact. 450 people worked at the plant. Although some operations remained on the site (e.g. warehousing, after-sales),

plant closure resulted in 337 job losses. Hoover-Candy's demise marked the end of major manufacturing in Merthyr Tydfil. The public sector dominates employment in the town (the council, the health service, the Welsh Assembly). In June 2012, 12.1% of Methyr Tydfil's workforce was unemployed (the second highest in Wales) and 7.1% of the adult population were benefit claimants.

Since 1999, lack of employment has led to a fall in the town's population. Merthyr is the only district of Wales where further depopulation is expected in future. By almost any measure — social well-being, health, educational attainment, wages, life expectancy, alcohol abuse, house prices — the town is struggling to survive.

Global networks

Link with wealth and poverty

Revised

Global networks

Global networks such as telecommunications (including the internet) and air transport facilitate flows of information, money, people and trade, which are the life blood of globalisation (Figure 2.3). In the modern world, people and places disconnected from these networks suffer major economic disadvantage.

Enquiry question: Why, as places and societies become more interconnected, do some places show extreme wealth and poverty?

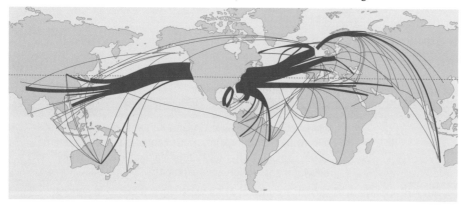

Traffic flows (Mbps)

5,000   2,500   1,500   100

Figure 2.3 Global telecommunications traffic

Network flows

Capital. Flows of capital include FDI, international aid, loans and **remittances** from migrants. Movements in one direction are often balanced by counter movements. Thus FDI generates profits that flow back to company headquarters and shareholders in MEDCs and emerging economies. Loans from **multilateral organisations** such as the IMF and World Bank have to be repaid with interest. Even international aid is often 'tied', providing significant economic benefits to donor countries.

International migration. Globalisation has increased the number of international migrants. Approximately 200 million people migrate internationally every year. In 2011, international migrants remitted over $US370 billion. The value of remittances to poor countries exceeds that of all development aid.

Remittances are the flow of money from international migrant workers back to their families in their home countries.

Typical mistake

The international migration of workers does not just involve low-skilled workers from Africa and south Asia migrating to the Gulf States, or Mexican workers migrating to the USA. It also includes highly skilled and highly educated workers from MEDCs working in global centres such as London, New York and Shanghai.

Exam practice answers and quick quizzes at **www.hodderplus.co.uk/myrevisionnotes**

International trade. International trade is overwhelmingly between MEDCs and within regions such as Europe and Asia. However, trade between rich countries and emerging economies is increasing rapidly. Europe accounted for 41% of international trade in 2010; Asia for 29%. Africa is least involved, with only 3% of world trade (Figure 2.4). Some of the world's poorest economies such as Burundi, the Central African Republic and Haiti are largely isolated from international trade, with exports comprising less than 10% of GDP. International trade is a key component of economic development — hence the promotion of free trade by the World Trade Organisation (WTO).

FDI. TNCs invest where returns are most certain. This is likely to mean places with high quality and/or low-cost workforces (human capital), physical resources, good accessibility, external economies, well-developed infrastructure and so on. As a result, global FDI flows are highly uneven (Table 2.3). Thus Africa, with nearly 10% of the world's population, received just 4.5% of total FDI in 2010. In contrast, Europe received one-third of all FDI, despite having only 6.3% of the global population. Most FDI flows are between rich countries and emerging economies. Capital flows are essential to 'switch-on' economic growth and development in LEDCs. Without it there is little chance of these countries escaping poverty.

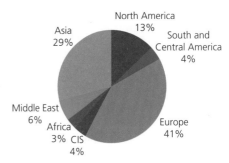

Figure 2.4 Distribution of international trade, 2010

Table 2.3 Regional population and FDI, 2010 (Source: UNCTAD)

	Global population (%)	Global FDI (%)
West/Central/Southern Europe	6.3	33.6
SE Europe/CIS	4.1	6.3
Africa	9.5	4.4
Latin America and Caribbean	5.3	12.5
Asia and Oceania	61.7	26.7
USA	2.8	16.5

Global networks and wealth and poverty

Communications networks, particularly telecoms, transform space, making geography and physical distance less relevant to economic activity. Transmission of information through global networks increases the influence of major world hubs such as London, New York, Tokyo, Singapore and Hong Kong as the command and control centres of the global economy.

Regions isolated and 'switched off' from global networks and new technologies range in scale from Sub-Saharan Africa to rural communities in northern Britain without access to broadband. These places, disconnected from the global economy, have limited attraction for inward investment and remain relatively poor. In contrast, many places with direct network access in western Europe, east Asia and the USA, 'switched in' to the global economy, have much greater opportunities for wealth generation and prosperity.

Now test yourself

11 What is meant by the term *remittances*?

12 Name three flows that are characteristic of globalisation.

Answers on p. 122

Tested

Examiner's tip

When revising global networks (a) identify the networks that play a key role in globalisation and (b) show how places (preferably named examples) connected to these networks are advantaged, and those disconnected are disadvantaged.

Now test yourself

13 Why are places that are well connected to global networks likely to be more prosperous than those that are poorly connected?

Answer on p. 122

Tested

The role of technology in a shrinking world

Revised

Advances in telecoms and air transport are the key to understanding the recent globalisation of the world economy. These advances have created a more interdependent and smaller world. They also play a crucial part in the emergence of a single global market, global sourcing of materials and labour, and **offshoring** the production of goods, services and research and development.

Offshoring describes the transfer abroad of manufacturing and services operated by TNCs to exploit lower costs. Offshoring makes a significant contribution to globalisation.

Telecoms and the internet

Information technology and the internet:

● allow people to interact, regardless of physical proximity
● are cheaper, faster and more efficient than previous forms of communication
● allow communication through e-mails to be delivered in seconds

- facilitate face-to-face meetings and decision-making (e.g. through technologies such as Skype and teleconferencing)
- enable TNCs to control production worldwide and respond quickly to market changes

Global financial services and retailing have been transformed since the development of the internet in 1995. Financial services have benefited from easier and instantaneous access to customers and to information about financial markets. Interaction between traders and the direct access to markets has improved efficiency and the speed of transactions. The City of London's advanced telecoms and information infrastructure has enabled it to maintain its position as the world's leading financial centre.

E-commerce has also had a major impact on retailing. In 2011, 12% of all retail sales in the UK were online. The leading players were Amazon, Argos and Tesco. With 82% of British adults using the internet in 2011 and 44% owning smart phones, the potential for marketing, selling and buying products and services online will continue to grow rapidly.

Air transport

Transcontinental and intra-continental air transport has become faster and cheaper, contributing to global interdependence. Aeroplanes have got bigger (e.g. A380, A330, Boeing 747) and more fuel efficient (Boeing 787). The Airbus 380 can accommodate more than 400 passengers and has a longer range than conventional airliners. Cities previously too far apart for direct flights (e.g. Dallas–Sydney, New York–Seoul) can now be reached non-stop, minimising travel time and maximising destination time for business transactions.

In a globalised economy, good connectivity with major airport hubs plays a leading part in the success of major world cities.

Sea transport

Falling unit costs of sea transport have boosted world trade and stimulated globalisation. Transport costs for freight, even when shipped half-way round the world, constitute only a tiny fraction of final prices. For example, transport costs for iron ore shipped from Australia to Europe are just US$12/tonne. Transport adds just 1.2% to the price of 1 kg of coffee, and a 20 tonne container can be freighted from east Asia to Europe for between US$600 and US$800.

Falling transport costs are the result of **economies of scale** resulting from the introduction of ever-larger container ships and oil tankers. The newest generation of Panamax container ships are nearly 400 m long with a payload of 11,000–14,500 containers. The largest supertankers transport nearly 500,000 tonnes of crude oil at a time. Lower transport costs and telecom technology have also made it cost-effective to transfer production from North America and Europe overseas, thousands of kilometres from markets.

Winners and losers

Revised

While some places are fully integrated into the global economy (i.e. 'winners') others (or 'losers') have hardly been touched. These differences owe much to **connectivity**, which in turn is a major factor in inward investment. Even so, the 'winners' usually have other **initial** or **acquired advantages** such as physical resources, human capital, political stability and good governance.

Physical resources

The success of emerging economies such as China and India has caused demand (and prices) for commodities such as oil, iron ore, copper and bauxite to soar in the past decade. China has scoured the world to secure vital mineral and

> **Typical mistake**
>
> Outsourcing and offshoring are often confused. Outsourcing describes the contracting-out of services and production by a business to other businesses. Offshoring is where work is done abroad, but still within the same business or a subsidiary.

> **Now test yourself**
>
> 14 What is:
> - (a) offshoring
> - (b) outsourcing
> - (c) e-commerce?
>
> Answers on p. 122
>
> Tested

> **Examiner's tip**
>
> Remember that cheaper and more efficient communication and transport networks are a major catalyst to globalisation. They have helped to create world markets, generate global competition and promote unprecedented levels of global interdependence.

> **Initial advantage** refers to the reasons for an economic activity to first locate at a place (e.g. energy, minerals, cheap labour).
>
> **Acquired advantages** are the economic benefits of long-established urban and industrial regions (e.g. clusters of linked activities and external economies of scale).

energy supplies. As a result, several resource-rich countries in SSA (e.g. Congo DR, Angola) have benefited economically. Thanks to Chinese investment in mining and transport infrastructure, the Congolese economy grew by an impressive 6.5% in 2011. This is despite an ongoing civil war, poor governance and the country's ranking as the world's least developed state. At least in the globalisation stakes, the Congo DR is a 'winner'.

Transport infrastructure

Brazil's modern Tubarão port complex is an example of the investment needed to connect producers with global markets. In 2010 Tubarão, connected to mines in the interior by a purpose-built 900 km railway, exported a record 104 million tonnes of iron ore. Of all exports from Tubarão, 15% go to China.

However, possession of mineral and energy reserves does not guarantee inward investment. Poor connectivity leaves many developing countries isolated from the global economy. Africa has 15 landlocked countries, some thousands of kilometres from the coast. Eight of these 'global backwaters' are among the 15 least developed countries of the world.

The seaboard of eastern China is the industrial powerhouse of the country's export-led economic growth. There successful manufacturing depends on good connectivity to **supply chains**. Although labour costs are significantly lower in interior China and in neighbouring Vietnam and the Philippines, the advantages of low labour costs are offset by the remoteness of these places from industrial **clusters** such as the Pearl River Delta and Guangdong, and poor access (via ports) to export markets.

Human capital

Human capital describes the education and skills of a country's workforce and is the main asset of the world's most advanced economies. A high-quality workforce is a prerequisite for success in a competitive globalised economy. Emerging economies such as China and India have large pools of highly skilled labour. China produces 75,000 graduates with higher degrees in engineering or computer science every year; India 60,000. While labour costs in both countries are rising quickly, unit costs (i.e. wage costs related to productivity) remain competitive and sustain economic growth. The result is a **virtuous circle** of rising real incomes, increasing consumption and expanding domestic markets.

In many parts of Africa and south Asia low literacy levels and skills are both a cause and an effect of poverty and underdevelopment. In LDCs the average length of schooling is often less than six months. Given these circumstances, even with modern telecoms, people lack the education and skills needed to work in offshored activities such as call centres and back office jobs.

Now test yourself

15 Labour costs are rising quickly in China, but unit labour costs remain relatively low. Why?

Answer on p. 122

Tested

Examiner's summary

Globalisation
- Globalisation is a complex process that drives economic and social change, but is also a summary description of these changes.
- TNCs, large companies with global production and marketing, may operate in either the manufacturing or service sectors.
- Inward investment by TNCs is often controversial, creating both advantages and disadvantages for recipient countries.
- Globalisation is not a new process; what is different about contemporary globalisation is its scale and its speed over the past two or three decades.
- International migrations are an important element of globalisation and involve the movement of both unskilled and highly skilled workers.
- Case studies of globalisation must focus on specific places, and refer to specific patterns, trends, histories and statistics.
- Economic development is a continuum and categories such as MEDC, LEDC and LDC are no more than convenient labels to assist generalisation.
- Efficient connections to global telecoms and transport networks are essential to economic growth and the development of nations.
- Cheaper and more advanced telecoms and transport systems are a catalyst to globalisation, but they are only one of several factors driving contemporary globalisation.

Roots

Population change using family records

United Kingdom

Huge changes have occurred to the UK's population over the past two centuries.

- There has been a massive increase in total population, from 10.5 million in 1801 to 63.1 million in 2012.

- There have been steep declines in **fertility**, family size and mortality. In the early nineteenth century the **crude birth rate** (CBR) was around 40 per 1,000, the **crude death rate** (CDR) 25 per 1,000, and the average family had five or six children. In 2010 the CBR and CDR were 11.6 and 7.9, respectively, and completed families had two children.

- The population is **ageing** because of falling fertility and mortality, and increases in life expectancy. In 1837 when official registration of births, marriages and deaths became compulsory, average life expectancy was around 38 years. By 2010, life expectancy had risen to 78 years for males, and 82 years for females.

- There is greater mobility, with increases in the frequency and volume of internal and international migration. In 2010, 591,000 immigrants entered the UK and 339,000 people emigrated.

Family histories and records of population change

Reconstructions of family histories in England and Wales rely on:

- censuses of population
- civil registration
- parish registers

Censuses

The first official census of population was held in England and Wales in 1801. Since 1801 censuses have been held at 10-year intervals (with the exception of 1941). The 1801 census was little more than a counting of heads. Detailed household returns are available for the censuses between 1841 and 1911. They provide information on the addresses of households, the names and ages of occupants, their gender, the relationship of occupants to the head of household (e.g. wife, son, daughter), place of birth, and occupation. Because of the 100-year secrecy rule, it is not possible to access household census records after 1911.

For family histories, census records allow reconstruction of:

- households, their size, age structure and changing composition
- the migration of households and individual household members
- the socio-economic status of households, from occupations and places of residence

Civil registration

Civil registration was introduced in 1837, making it a legal obligation to register all births, marriages and deaths. Birth certificates record the name of a child, names of parents and father's occupation. Marriage certificates give details of date of marriage, the ages, occupations and residences of the bride and groom, as well as providing the names and occupations of their fathers. Death certificates provide additional information such as the place and cause of death.

Civil registration provides accurate records that allow the analysis of fertility, mortality and marriage at family, local, regional and national scales.

Parish registers

Before civil registration, family records rely on parish registers that record baptisms, burials and marriages. Their accuracy is variable, depending on the

---

Enquiry question: How does evidence from personal, local and national sources help us understand the pattern of population change in the UK?

Examiner's tip

The CBR and CDR do not measure fertility and mortality accurately. As simple ratios of the number of births and deaths per 1,000 of the population they are strongly influenced by the age–sex structure of a population.

Now test yourself

16 What is the difference between the crude birth rate and fertility?

Answer on p. 123

Tested

Now test yourself

17 What demographic information is available from civil registration?

Answer on p. 123

Tested

Examiner's tip

It is instructive to study census returns, civil registration and parish registers as primary documents to investigate your own family history or local demography. Today this is done most easily online at sites such as:
www.parishregister.co.uk,
www.ancestry.co.uk, and
www.findmypast.co.uk

conscientiousness of parish priests and the number of non-Anglicans resident in the parish. Closed rural societies tend to have more accurate parish records than urban ones, where migration was greater. There is also a clear decline in the accuracy and completeness of registers with the onset of industrialisation and urbanisation in the late eighteenth century. Most original parish registers are kept in County Record Offices. Many have been transcribed, some are available on microfilm and CDs, and some online.

Table 2.4 Possible investigative studies of personal and local population change

Source	Family studies	Local studies
Census data (1841–1911)	Family size Population movements Birth intervals Occupations and socio-economic status	Population change Changing age–sex structure Changing migration patterns Changing employment patterns
Civil registration data	Family size and fertility Age at marriage Mortality and life expectancy Infant mortality Occupational structure Literacy	Changing CBRs and CDRs Natural population change Changing fertility and mortality Changing marriage rates
Parish register data	Family size and fertility Birth intervals Life expectancy	Trends in baptisms and burials Natural population change Age at marriage and marriage prevalence Employment patterns

Trends in the UK population since 1900
Revised

The UK's population has undergone significant changes since 1900. These changes are the result of trends in births, deaths and migration.

Births

The number of births in a population is determined by:

- **fertility** (i.e. the average number of children in completed families, known as the total fertility rate (TFR))
- the number of women of reproductive age (i.e. 15–44 years)

Average fertility in the period 1901–1905 was 110 live births per 1,000 women aged 15–44. By 2009 this figure had almost halved to 63.6. In 1901 the TFR was 3.5 children, compared with 2 children in 2010. The general trend of fertility decline has been interrupted three times: in the decade after the Second World War (the so-called post-war 'baby boom'), in the 1960s, and most recently since 2008 (Figure 2.5).

Typical mistake

In demographic terms *fertility* refers to the average number of children born to each woman either in her lifetime or within specific age groups. Don't confuse this with *fecundity*, which is the ability to have children.

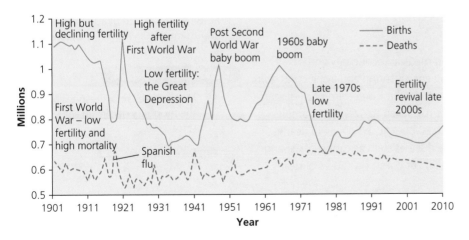

Figure 2.5 UK births and deaths 1901–2010

Causes of fertility decline

The decline in fertility during the twentieth century is largely due to social and economic change:

- greater social and economic equality between men and women
- later average ages of marriage
- advances in reducing childhood and infant mortality, resulting in the vast majority of children surviving to adulthood
- extending the period of full-time education, increasing dependency and making children less of an economic 'asset'
- increased availability and the social/moral acceptability of artificial contraception; oral contraceptives became widely available in the 1960s, giving women greater control of their own fertility

The fertility peak that occurred after 1945 and lasted until the mid-1950s was the result of postponement of marriages during the Second World War. The rise in fertility since 2008 has been driven by high levels of international immigration. Thus while British-born women in 2009 had TFRs of just 1.68 the TFRs of some immigrant groups from Africa and Asia were three times greater.

Deaths

Mortality, like fertility, declined for most of the twentieth century. The CDR fell from an average of 16 per 1,000 in 1901–2005, to 9.2 in 2006–2010. The decline in infant mortality was particularly dramatic: from 140 per 1,000 in 1900, to 4.35 in 2010 (Figure 2.6). As mortality declined, life expectancy increased. In 1901 life expectancy at birth averaged 45 years for men and 49 years for women. By comparison in 2011 it was 78 years for men and 82 years for women.

The First and Second World Wars and the Spanish influenza pandemic of 1919 caused sharp, but temporary increases in mortality. Around 750,000 young men died in the First World War (contributing to the fall in fertility in the 1920s), and Spanish flu is estimated to have killed 150,000 in the UK.

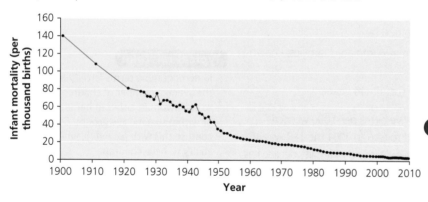

Figure 2.6 Infant mortality decline, UK 1901–2010

Causes of mortality decline

Declining mortality was due to advances in medical technology and improvements in healthcare, sanitation, hygiene, nutrition and living conditions. In 1880 infectious and parasitic diseases accounted for one-third of all deaths. By the early twenty-first century, this figure had fallen to 17%. For example, tuberculosis killed killed 80,000 people in 1880 but caused only 440 deaths in 1997. As people live longer, degenerative diseases such as cancer (43%) and circulatory diseases (26%) have become the most common causes of death.

International migration

Since the early 1990s, migration has been the main driver of population growth in the UK, with immigration consistently exceeding emigration. Half of all immigrants have come from New Commonwealth countries (mainly in south

Examiner's tip

Avoid sweeping generalisations when explaining fertility, mortality and immigration trends. Remember that large populations are highly differentiated along social, economic, cultural and demographic lines.

Now test yourself

18 Give reasons for the post-1945 surge in fertility in the UK.

Answer on p. 123

Tested

Typical mistake

It is easy to overstate the importance of medical technology in the decline of mortality in the twentieth century. Improvements in environmental conditions (sanitation, hygiene, housing) and nutrition also played an important part in mortality decline.

Now test yourself

19 Name three factors responsible for the mortality decline in the UK since 1900.

Answer on p. 123

Tested

Asia) and one-third from Europe. The enlargement of the EU in 2004 resulted in large influxes of migrants from eastern Europe, with the UK recording its highest ever levels of net migration gain (approximately 250,000 per year) (Figure 2.7).

Typical mistake

The large net migration gains experienced by the UK since 2000 are inflated by thousands of foreign students, most of whom will not settle permanently in the country.

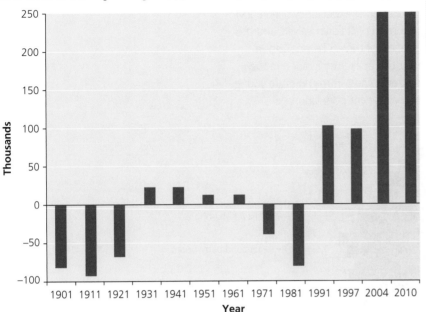

Figure 2.7 Net international migration, UK 1901–2010

Causes of immigration

Economic **push** and **pull factors**, such as the prospect of a higher standard of living and better quality of life, largely explain immigration flows from the New Commonwealth to the UK. Having settled in the UK, legislation allowed single migrants to be joined by spouses and their dependants.

Successive UK governments have encouraged immigration because most in-migrants are young adults and are economically active.

In 2004, with the expansion of the EU, EU8 citizens acquired legal rights to live and work in the UK. Between 2004 and 2010 1.9 million immigrants migrated to the UK from eastern Europe. They provided a source of skilled and relatively cheap labour that the British economy needed in the economic boom years of 2004–2007. For immigrants the main attraction was the UK's high wage economy.

In the decade 2000–2009 many immigrants entered the UK as political refugees or asylum seekers, fleeing war and persecution in the developing world. Applications for asylum peaked at 84,000 in 2002; by 2011 they were down to 23,000.

Push and **pull factors** explain the movement of migrants. Push factors are the negative features in the migrant's place of origin (e.g. unemployment). Pull factors are the attractions of a migrant's destination (e.g. political stability).

Typical mistake

It is wrong to think of international immigration to the UK as solely comprising New Commonwealth immigrants. There is also a significant migration stream to the UK of highly qualified, professional workers essential to globalised service industries (e.g. banking) and the NHS.

Population growth

The UK's population increased from around 38 million in 1900, to just over 62 million in 2011 (Figure 2.8). The fastest growth — 10% — occurred in the decade 1901–1911; the slowest — 0.7% — was between 1971 and 1981. In only one year — 1976 — did deaths outnumber births to give **natural population decrease**. The contribution of immigration to population growth in the UK was highest in the period 2004–2012. In 2010 natural increase accounted for nearly 40% of population growth; the rest was due to immigration.

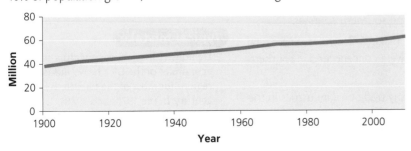

Figure 2.8 Population change in the UK, 1901–2011

Age structure

Since 1900, changing patterns of fertility, mortality and migration have modified the UK's age structure. The most obvious change has been the ageing of population because of declining fertility and increasing life expectancy (Figure 2.9). In the course of the twentieth century the proportion of children halved, and the proportion of over 64 year olds increased by a factor of three. The population pyramid therefore aged, both at the 'base' and at the 'apex'. Immigration in the past decade has temporarily slowed the ageing trend. Most immigrants are young adults with higher birth rates and fertility than the resident population.

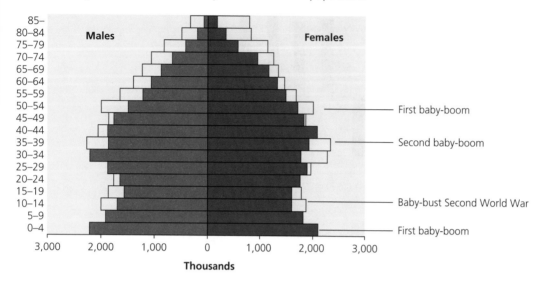

Figure 2.9 Changing age structure in the UK, 1951 and 2001

Internal migration

Changes to population distribution within the UK are largely explained by internal migration. Since 1900 there have been four major trends in internal migration:

- at the national level a general north–south drift of population from Scotland, northern England and Wales to London and the southeast region.

At the regional level:

- the **suburbanisation** of populations in large towns and cities
- **counterurbanisation**, and population growth of small towns and rural areas, most often within commuting distance of large urban centres
- retirement migration from cities and conurbations to regions of recreational and high amenity value

North–south drift

A north–south drift of population, driven by economic factors, prevailed for most of the twentieth century.

- By 1919 many heavy industries such as coal mining, steel, textiles and shipbuilding in central Scotland, northern England and south Wales were already in decline. This decline accelerated during the Great Depression of the 1930s. Thousands of workers and their families, impoverished by long-term unemployment, migrated to London and southern counties.
- New, dynamic growth industries such as car making, electronics and aircraft manufacture developed in the 1930s. Most were concentrated in southern Britain.
- **Deindustrialisation** and the collapse of many traditional industries in the 1970s and 1980s. These industries, often undercapitalised, overstaffed, heavily unionised and with inflexible workforces, could not compete with more efficient and lower-cost producers in Asia.

> **Typical mistake**
>
> Deindustrialisation was a feature of the economies of the UK's heavy industrial regions for most of the twentieth century. In the 1970s and 1980s, deindustrialisation simply accelerated long-term industrial decline.

Suburbanisation

The suburbs of most British towns and cities grew rapidly during the inter-war and immediate post-war years. For example, between 1921 and 1951, Outer London's population doubled to 4.5 million. At the same time the population of Inner London fell from 5 million to 3.7 million (Figure 2.10). Improvements in transport (e.g. extensions to London's underground, buses and, later, private car ownership) made suburbanisation possible. Suburbanisation also reflected rising living standards and aspirations.

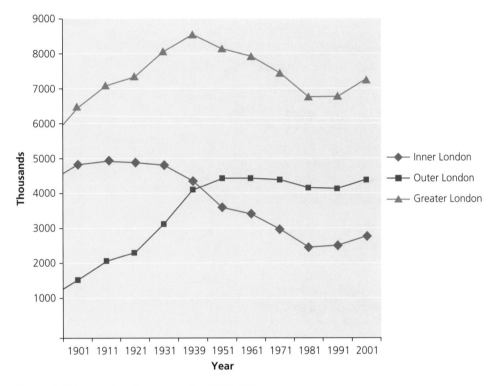

Figure 2.10 Population change, London 1901–2008

Counterurbanisation

Counterurbanisation is the urban to rural shift of population. In the UK it began in the 1970s and continues (though less strongly) today. It has resulted in a small increase in the percentage of rural dwellers, reversing the urbanisation trend that dominated the previous 200 years. By 2001–2010 counterurbanisation had lost much of its impetus: the UK's rural population grew only marginally faster than the urban population. The gap between rural and urban population growth narrowed because international immigrants mainly settled in towns and cities.

Most urban–rural migrants are in the 45–65 year age group, work in nearby towns and cities and can afford the extra costs of housing and commuting. The need to live within commuting distance of workplaces explains the rapid population growth in many rural counties adjacent to cities and conurbations (e.g. Warwickshire, Cheshire, North Yorkshire). However, thanks to the internet and modern telecoms, a larger proportion of rural dwellers (compared with urban dwellers) are able to work from home and are self-employed.

A combination of **push** and **pull factors** explain counterurbanisation. Push factors include high crime rates, antisocial behaviour, pollution, traffic congestion, lack of community, a rundown physical environment and poor services in large towns and cities. In the UK, the attraction of the countryside (the rural 'idyll') is particularly strong and reflects cultural values that are often quite different from those of other, more urbanised European cultures.

> **Typical mistake**
>
> It should be emphasised that counterurbanisation implies some redistribution of population from urban to rural areas, resulting in rural population growing faster than urban population. Hence the proportion of a population living in rural areas increases.

> **Now test yourself**
>
> 20 What are 'push' and 'pull' factors?
>
> **Answer on p. 123**
>
> Tested ☐

Retirement migration

Counterurbanisation is not just about commuters. Increasing numbers of older people (baby boomers), with generous pensions and/or valuable properties in urban areas, move to rural areas on retirement. Among the most popular destinations are coastal resorts such as Christchurch (Dorset) and Sidmouth (Devon) and historic market centres such as Taunton (Somerset) and Dorchester (Dorset).

The UK's greying population

Revised

Low and declining fertility together with increasing life expectancy are responsible for the ageing of the UK's population — a trend shared with most other developed countries.

Between 1985 and 2010, the median age of the UK population increased from 35.4 to 39.7 years, and the proportion aged 65 years and above rose from 15% to 17%. By 2035, nearly one-quarter of the population will be aged 65 years or more. This steep increase is explained by the 'baby boomers' of the post-war years and 1960s entering retirement. However, the fastest-growing cohort of over 65s is the 85+ group. This group doubled in size between 1985 and 2010 and is expected to represent 5% of the population by 2035.

Economic and social effects of ageing

Economic

Ageing causes economic problems because most people aged 65 and over are retired, and are consumers rather than producers. Moreover, this group is increasing faster than the number of adults in work.

Ageing causes economic problems because:

- in the UK people aged 65 and over receive state pensions paid out of taxation
- elderly people also make heavy demands on medical and healthcare services
- numbers of extreme elderly (aged 85 and over) are growing rapidly (hospital and community care for the over 85s is three times greater than for 65–74 year olds)

In total 65% of all expenditure by the Department of Work and Pensions goes to those above working age, while average public spending on retired households is twice as great as on non-retired.

Social

Older people are more likely to live alone than any other group. This leads to isolation, loneliness and social exclusion. As more people live to extreme old age, and physical and mental impairment increase with age, support either from families or the state is essential. The social problems of ageing are compounded by modern families becoming more dispersed geographically, and the decline of extended families (i.e. containing several generations) that cared for elderly relatives in the past.

Solutions

Governments are under growing pressure to tackle the problems of ageing populations. Policy options include increasing taxes, promoting immigration, re-structuring occupational pensions, raising the age limit for state pensions, encouraging old people to work longer, and elderly owner-occupiers to sell

> **Examiner's tip**
>
> Remember that in the absence of large-scale immigration, the inevitable consequence of prolonged fertility and mortality decline is ageing of a population and increased dependency. The outcome is economic and social problems, which most MEDCs have yet to address.

their homes to pay for residential care and nursing homes. Some of these options are outlined in more detail below.

- The burden of additional taxation impacts most on those who are economically active and is unpopular. Thus democratically elected governments are often reluctant to choose this option.
- Most immigrants are young adults who are in work, pay taxes and help to reduce dependency. In 1985 the UK had the second most-aged population in Europe. Thanks to large-scale immigration since the late 1990s, the UK now ranks seventeenth.
- Government and business acknowledge that final salary pension schemes are unaffordable. Gradually they are being replaced by cheaper alternatives that rely on investments and the performance of the stock market.
- Pensionable age in the UK is being raised in line with life expectancy. In 2020 the state pension age for men and women will rise from 65 to 66 years, and in 2026 to 67 years. Further increases are expected after 2026.
- Recent legislation prevents employers forcing workers to retire when they reach 65 years of age. This means that many older people will continue to work and pay taxes. Already 1.8 million people aged 65 and over are in employment in the UK.

On the move

Key migrations into Europe
Revised

Post-colonial immigration

In the 1950s and 1960s many European governments encouraged short-term immigration to relieve labour shortages in their economies. The main post-colonial migration flows into Europe at this time were immigrants from British, French and Portuguese colonies in Sub-Saharan Africa, North Africa, the Caribbean and south Asia. Also significant was the movement of Turkish migrants to post-war Germany. Immigrants had a number of features in common:

> **Enquiry question**: How is migration changing the face of the EU?

- they were mainly single men, employed in unskilled work in industries such as textiles and public services
- as economic migrants they saved and **remitted** money to families and relatives in their countries of origin
- having accumulated sufficient capital most intended to return home to their families overseas

Immigration flows after 1970

After 1970 the main features of immigration into Europe were:

- the continuing importance of immigration from less developed regions in Asia and Africa (Figure 2.11)
- huge increases in the total number of immigrants (including significant numbers of illegal immigrants)
- immigration for social as well as economic reasons, i.e. the migration of families to settle permanently and join single (usually male) family members — a process known as **family reunification**
- large increases in the numbers of political refugees or **asylum seekers** fleeing wars and persecution in countries such as Iraq, Iran, Afghanistan, Sri Lanka, Somalia and Zimbabwe

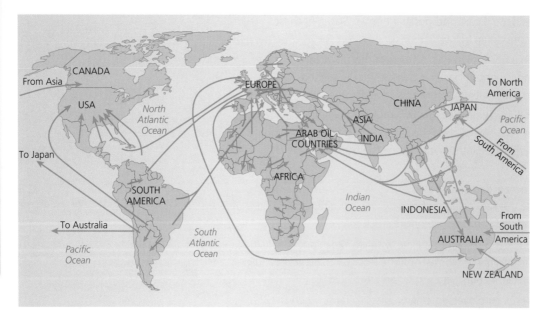

Figure 2.11 International migration flows

Now test yourself | Tested ☐

Now test yourself

21 In the context of migration, what is meant by the term 'family reunification'?

22 Why did the number of asylum seekers in western Europe rise steeply in the early twenty-first century?

Answers on p. 123

Examiner's tip

Migrations can be classified in a number of ways, which provide a useful framework for exam answers. Among the classifications, most widely used are: international/intranational, rural–urban/urban–rural/urban–urban, voluntary/forced, economic/social, single person/family.

Case study UK

Post-colonial period. Large-scale international immigration to the UK from outside Europe began after the Second World War. People from the Commonwealth and colonies with British passports were legally entitled to settle in the UK. The first wave of immigrants came from the West Indies in the 1950s to mid-1960s. They mainly settled in London and other major cities such as Birmingham, Bristol and Leeds. Large numbers of Pakistanis also gravitated to smaller northern industrial towns such as Oldham, Burnley and Keighley, meeting the demand for 24-hour labour in the textile mills.

South Asian migrants began arriving in large numbers from the mid-1960s: first Indians and Pakistanis and later Bangladeshis. Young Pakistani men often migrated with close male relatives. The majority came from just a few districts in rural Pakistan, such as Mirpur in Kashmir. They were economic migrants. Wages for low-skilled work in the UK were 30 times greater than in Pakistan and cheap flights made it possible to retain contact with their families overseas.

After 1970. The government, responding to popular concerns about the scale of international immigration in the 1970s, placed restrictions on entry to the UK. British passport holders born overseas could only settle permanently in Britain if they (1) had a work permit or (2) could prove that a parent or grandparent had been born in the UK.

From 1970s onwards, the nature of immigration from south Asia began to change. Many single male immigrants settled permanently and were joined by their spouses and extended families. In addition, second-generation immigrants often sought marriage partners in Pakistan and India. Marriage migration has been maintained, with 60% of marriages of British-born Pakistanis marrying partners born in Pakistan. This practice has been strengthened by cultural preferences for marriages within families, i.e. first cousin marriages. Despite further attempts to curb immigration from outside the EU, between 2002 and 2011 1.8 million migrants from Commonwealth countries entered the UK.

The UK's black African population increased between 2001 and 2011, passing 900,000 in 2011. Primarily located in London, many black African immigrants are asylum seekers, escaping war and persecution in countries such as Somalia, Zimbabwe, Congo DR and Eritrea. Political persecution was also responsible for the sudden influx of 60,000 Asians of Indian descent, expelled from Uganda in 1973.

Table 2.5 Ethnic minority groups in the UK, 2011 estimates

	Population (millions)	Annual growth (% 2001–2011)
Indian	1.41	3.7
Pakistani	1.12	5.8
Bangladeshi	0.447	6.3
Afro-Caribbean	0.595	0.6
Black African	0.990	10.8

Germany

Since the 1950s Germany's spectacular economic growth has relied heavily on immigrant workers. Unlike the UK and France, Germany could not turn to a former empire and colonies as a source of cheap immigrant labour. Initially, most immigrant workers came from southern Europe, but by the early 1960s Turkey became the main source.

The nature of immigration was comparable with other European countries. Most migrants were single men and provided unskilled labour. Immigrants were not encouraged to settle permanently, hence their popular name 'guest workers' or *gastarbeiter*.

From around 7,000 in 1961 the number of Turks working in Germany rose to more than 900,000 in 1973. Today there are 2.5 million people of Turkish origin living in Germany, making them by far the largest immigrant group. They account for 60% of Germany's Muslim population.

Gradually, Turkish immigration has become permanent. Immigration trends in the past three decades have been similar to those in the UK and France, with Turkish immigration increasingly focused on family reunification, and young Turkish men with German citizenship returning to Turkey to find marriage partners.

Typical mistake

Given the negative publicity that international immigration generates, the fact that many European countries have encouraged immigration over the past 50 years because of its economic benefits (e.g. skills, cheaper labour, more youthful population) is too often overlooked.

Examiner's tip

It is useful to think of the growth of international migration, and increasing multicultural nature of societies in MEDCs, as an inevitable consequence of the globalisation of the world economy.

Key migrations within Europe

Revised

Case study Immigration to the UK from eastern Europe: 2004–2011

Scale. The scale of immigration from Europe to the UK between 2004 and 2010 was unprecedented. European immigration averaged 170,000 per year — almost three times greater than for the period 1997–2003. By 2011, 950,000 people born in **EU8** countries were resident in the UK, plus 134,000 from Romania and Bulgaria (EU2) (Figure 2.12).

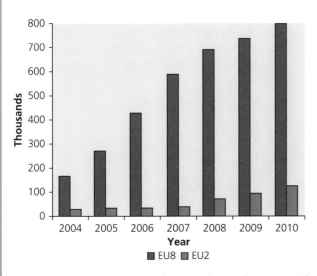

Figure 2.12 EU accession population in the UK (Source: ONS)

Causes. The stimulus to migration was the accession of the EU8 countries to the EU in 2004 when EU8 citizens were granted full and immediate access to the UK's labour market. Migrants from eastern Europe, and in particular from Poland, flooded into the UK. The scale of immigration was exceptional. This was partly because most EU15 countries restricted immigration from the EU8 for a transitional period until 2011.

Immigration was driven by economic factors. In 2007, GNI per capita in Poland was just US$9,850, compared with US$40,660 in the UK. Unemployment was twice as high in Poland, and the minimum wage was one-fifth of that in the UK. Although the wealth gap between the two countries had narrowed by 2012, it remained significant. East European immigration

peaked in 2007, with the recession after 2008 causing a steady decline in numbers as many migrants returned home. However, there was still a strong net inflow from the EU8 to the UK in 2011–2012.

Employment. By 2007, EU8 citizens represented 46% of all foreign immigrants working in the UK. The British government encouraged immigration because of its economic advantages:

- workers with skills needed to meet labour shortages during the 2004–2007 economic boom
- young, economically active immigrants helping to counter the economic and social problems caused by the UK's ageing population
- the support of businesses that favoured east European workers because of their willingness to take jobs (often low paid) rejected by the indigenous workforce, and their positive work ethic

The most popular employment sectors for east Europeans are hospitality and catering (18%), agriculture (10%), manufacturing industry (7%) and food processing (6%). In agriculture they account for 40% of all employment in the industry; in hotel and catering 10%.

Geography. Because east European immigrants are employed in relatively few economic sectors, their geographical distribution within the UK is highly localised. The main concentrations are in London, the Southeast and East of England. In rural areas with labour-intensive farming and food-processing industries such as south Lincolnshire, Cambridgeshire and Herefordshire they account for up to one in four of the workforce. London, with its huge hospitality and catering sector, has the largest concentration of east European workers.

EU8 refers to the eight accession countries that joined the EU in 2004 i.e. Poland, Czech Republic, Slovakia, Hungary, Slovenia, Estonia, Latvia and Lithuania (EU8 excludes Cyprus and Malta that also joined the EU in 2004). EU2 comprises Romania and Bulgaria that joined the EU in 2007.

Typical mistake

It is a mistake to think that all migrations are permanent. A large proportion of migrants aim to remit or save money and then return home; others may be forced to return having failed to qualify for permanent residence. Remember that for every migration there is always a counter-migration movement in the opposite direction.

Examiner's tip

The simplest way to revise the causes of migration, and structure your exam answers on the topic, is to think in terms of push and pull factors.

Case study Retirement migration flows to Mediterranean locations

Scale and character. Retirement migration from northern Europe to sunbelt locations in the Mediterranean has grown strongly in recent decades (Figure 2.13). Most migrants are Britons, Germans and Scandinavians. They favour coastal destinations in regions such as Andalucia, Alicante and the Balearic Islands in Spain; Tuscany in Italy; and the Algarve in Portugal.

Most migrants are:

- in their 50s or early 60s
- couples without dependent children
- relatively well-off owner-occupiers with at least one occupational or private pension

Spain is the preferred destination for British and German migrants, followed by France and Italy. Around 450,000 Britons, and one-quarter of all British pensioners living abroad, are in Spain. One-fifth of UK ex-pats in Spain are aged 65 years and over and 63% are either retired or economically inactive.

Causes. The rapid growth of retirement migration to the Mediterranean is due to:

- rising affluence among the populations of northern Europe
- increases in home ownership making people more mobile and, in countries with high property prices, creating substantial capital assets

- lower property prices in Spain and other Mediterranean countries compared with northern Europe
- earlier retirement and longer life expectancy, with most retirees anticipating 20 years or more of good health following retirement
- greater awareness of foreign places and cultures through overseas holidays and travel abroad
- cheap air travel, with budget airlines operating services from northern Europe to small, regional airports in southern Europe
- the creation of the EU single market removing barriers to movement, work and residence for EU citizens

Reasons for migration. The main reason for migration is the warm, sunny climate of the Mediterranean. Many migrants regard the climate as adding to the quality of life, providing opportunities for year-round outdoor recreation and health advantages. Economic reasons for migration include lower house prices and lower cost of living.

Many migrants retain properties in the UK and Germany, with **circular** and **seasonal** migration between the Mediterranean and northern Europe being common. In Spain, property developers, builders and estate agents have encouraged immigration by developing new settlements or *urbanizaciónes* to accommodate retirees, other permanent migrants and holiday home owners.

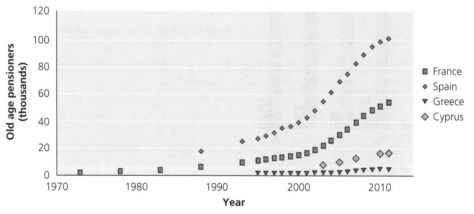

Figure 2.13 UK OAPs in France, Greece, Spain and Cyprus

Examiner's tip

Revision of case studies of migration should consider the scale and patterns of migration, the causes of migration and their economic, social and environmental impacts. Such studies must include specific details that locate migrations in space and time.

Now test yourself

23 Explain the growth of international migration on retirement in Europe.

Answer on p. 123

Tested

The consequences of migrations

Revised

International migrations into and internal migrations within Europe have important economic, social, environmental and political consequences.

Economic

Benefits

- Most international immigrants are young adults and taxpayers of working age.
- The majority of immigrants from North Africa, south Asia and Turkey take low-paid but essential jobs that are often unattractive to the indigenous workforce. The UK, Germany, France and other EU countries derive similar advantages from migrant workers from eastern Europe who are often better qualified, more productive and have a stronger work ethic than indigenous workers.
- Smaller immigrant streams, mainly from the USA, Australia and EU countries to the UK, comprise highly skilled, highly qualified and highly paid workers. In the UK their employment in international producer services (banking, law, advertising, etc.) is vital to the economy.
- In the UK there are nearly 300,000 foreign students from outside the EU in higher education. They contribute around £5.4 billion a year to the UK economy and have become essential to the budgets of many UK universities.
- Retirement migration can have significant economic benefits for destination areas such as southern Spain. Retirees are consumers using local services and paying taxes.

Now test yourself

24 What are the economic benefits of international immigration?

Answer on p. 123

Tested

Costs

Large-scale immigration also brings costs to receiving countries.

- Because most immigrants are young adults, they are more likely to have children than the indigenous population. Large geographical concentrations of immigrants place enormous pressure on schools, hospitals and other public services. Similar pressures on public services are found in inner cities with large clusters of Pakistani, Bangladeshi and African groups. This is due to high fertility as well as youthful population structures.
- Large influxes of migrants and their families exacerbate problems of housing shortages. Currently in the UK 40% of housing demand is due to immigration. The situation is most acute in London and the southeast, where a shortfall of 325,000 homes is forecast by 2025.
- In the Mediterranean region, large numbers of retirees from Europe make disproportionate demands on medical and health services.

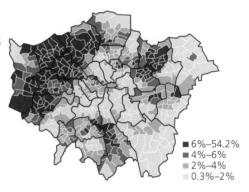

6%–54.2%
4%–6%
2%–4%
0.3%–2%

Figure 2.14 Spatial segregation: London's Indian population

Social

Whereas the economic benefits of immigration generally outweigh the costs, the social advantages are less clear cut.

Isolation, both social and spatial, is common among immigrant groups when the cultural gap between them and the host society is large. The outcome is **segregation**. Many ethnic minority groups cluster in urban enclaves (or **ghettoes**), often with minimal social contact and interaction with the host society (Figure 2.14).

Despite the long history of immigration into the UK from south Asia, segregation in British cities remains strong. This explained by:

- self-segregation and a desire on the part of some communities to maintain their own culture, language and traditions
- poverty, which forces many ethnic minority groups to cluster in cheap housing areas in inner cities
- discrimination and perceived threats from the host society

Typical mistake

Remember that a ghetto is a geographical concentration of particular socio-economic/ethnic/national groups in a town or city. Slum, on the other hand, refers to a run-down, decaying residential area with sub-standard housing and services.

Examiner's tip

Remember that immigration is a sensitive issue. Any assessment of the economic, social, environmental and political impact of immigration must be (a) balanced and (b) avoid any terms or arguments that might be construed as racist.

- government policies of multiculturalism, which have promoted the cultural differences and separate identities of ethnic groups at the expense of national cohesion
- Muslim and Hindu traditions of marrying partners from their country of origin, reinforcing cultural separation

Poverty, unemployment and social exclusion among ethnic minority groups can create a sense of injustice, which can occasionally erupt in public disorder (e.g. riots in the UK in 1980, 2001 and 2011).

Problems of integration are also found among British ex-pat groups in Spain. They too form enclaves of English speakers, with limited knowledge of the local language and little interaction with local people.

Now test yourself

25 Outline the social costs of large-scale immigration.

Answer on p. 123

Tested

Environmental

The environmental impact of immigration is largely negative. In the period 2001–2011 the UK recorded its most rapid decennial increase in population for over 200 years. Largely a result of immigration, the total population grew by 7% to 63.1 million.

In London and the southeast, already overcrowded, population growth could reach sustainable limits. Water shortages have already led to the construction of London's first desalination plant. Housing shortages put pressure on the green belt, resulting in loss of countryside, longer commuting times and more pollution. In Spain, immigration helped to fuel a building boom in the 1990s and early 2000s. Massive urbanisation (much of it often illegal) occurred on the Costa del Sol and other coastal regions with severe environmental impacts. Retirement migration to the Mediterranean also contributed to rising demand for scarce water resources for swimming pools and golf courses, which in the long term is unsustainable.

Political

Immigration, particularly from outside Europe, is controversial and has become highly politicised. At the 2012 French presidential election the right-wing *Front National* party, which is anti-immigration, won 18% of the popular vote. Politicians have responded to public unease about the scale of immigration. In France, popular concerns that North African immigrants fail to integrate with the French way of life have prompted the government to ban Muslim women and girls wearing the *hijab* in schools and universities and the *niqab* and *burqa* in public. Similar measures have been introduced in other European countries, notably in The Netherlands, Belgium, Spain and Switzerland. Meanwhile, immigration controls have been tightened in the UK (Table 2.6).

Table 2.6 Non-EU immigration controls in the UK

Income thresholds	The UK government will reduce immigrants' dependence on public funds. In 2012 new income thresholds will mean that a couple will need a gross income of between £18,600 and £25,700 to qualify for residence. Currently around 45% of immigrants meet this target.
Labour skills	In 2008 a points system was introduced and immigrants placed in five categories or tiers according to age, qualifications, past earning, etc. The aim is to coordinate immigration with needs of the economy.
Language skills	Spouses joining partners in the UK now have to show basic knowledge of English before being allowed to stay. Foreign students must have advanced English language skills.
Citizenship tests	A compulsory 45-minute test on British citizenship has to be taken by all immigrants from outside the EU.
Student visas	Student visas have been used as a means of illegal immigration. Visas are only sanctioned for higher education establishments, undergraduates can no longer bring family members with them to the UK, and the optional 2-year stay in the UK after graduation has been abolished.
Bogus marriages	Marriage partners must be aged 21 years and over, be married and plan to live together, and be able to support themselves without resort to public funds.

Examiner's summary

Population migration

✔ Population change can be studied at scales from the international and national, to the local and family.

✔ Census records, civil registration and parish registers are primary documents that can be used to reconstruct family histories.

✔ Crude birth and crude death rates are strongly influenced by age structure and are not therefore accurate measures of fertility and mortality.

✔ Declining mortality in MEDCs since 1900 is due to a range of factors: improvements in diet, sanitation, housing, etc. and not just to medical advancements.

✔ International immigration has both costs and benefits for sending and receiving nations.

✔ Counterurbanisation is the opposite of urbanisation and describes an increase in the *proportion* of rural dwellers in a country or region.

✔ Most MEDCs have still to tackle the economic and social problems caused by ageing populations.

✔ Increasing levels of international migration are simply a part of globalisation and the integration of the world economy. One inevitable outcome is that societies become more multicultural.

✔ For every migration stream there is always a counterflow in the opposite direction.

✔ Case studies of migration should consider scale, direction, causes and the economic, social, environmental and political impacts.

✔ Exam questions requiring assessment of the costs and benefits of immigration must deal with the topic sensitively. Answers must be fair and balanced and there should be no suggestion of prejudice nor any implicit racism.

World cities

Rural–urban migration

Revised

Massive **urbanisation** on a global scale has occurred in the past 60 years.

- In 1950 just under 30% of the world's population were urban dwellers. Today the proportion is 52%.

- Most of this increase has taken place in the developing world where rates of urbanisation show no signs of slowing.

- By 2020 half of Asia's population will live in urban areas and Africa's population will be predominantly urban by 2035.

- Between 2011 and 2050 the total number of people living in towns and cities is expected to rise by 2.6 billion, and by mid-century will represent 70% of the global population.

> **Enquiry question**: What is driving the new urbanisation taking place and what are its consequences?

> **Urbanisation** is an increase in the proportion of urban dwellers in a country or region.

Million cities and mega cities

A feature of the new urbanisation is the increasing importance of large cities. Cities with more than 1 million inhabitants account for a growing proportion of the world's urban population — a trend that will continue in future.

Million cities have populations of 1 million and over. **Mega cities** are the world's largest cities, with 10 million or more inhabitants. Even larger are so-called **meta cities** such as Tokyo and Delhi, with populations in excess of 20 million.

> **Examiner's tip**
>
> Any definition of the term *mega city* is arbitrary; however, the most widely used definition is 'an urban agglomeration of 10 million people or more' (UN).

Table 2.7 Total urban population by city size (millions) (Source: UN)

Size	2011		2025	
	Population	Per cent	Population	Per cent
<1 million	2,214	61	2,482	53.5
1–5 million	776	21.4	1,129	24.3
5–10 million	283	7.6	402	8.7
>10 million	359	10	630	13.5
Total	3,632	100	4,643	100

Table 2.7 shows that both the number and proportion of urban dwellers living in million cities and mega cities will increase in future. In 1950 there were only two

mega cities: New York and Tokyo. By 2011 there were 23. Fourteen new mega cities will appear by 2035; 12 of these cities will be in the developing world. Their combined population will be 630 million and they will be home to 1 in 13 of the world's population.

The growth of million cities and mega cities

Two demographic factors influence the growth of million cities and mega cities: natural population increase and internal migration.

Natural population increase

Urbanisation is partly due to the excess of births over deaths in towns and cities.

● Urban crude birth rates (CBRs) are often high in LEDCs because rural–urban migration is age-selective and urban populations have large proportions of young adults. Compared with rural areas, urban crude death rates (CDRs) are relatively low. Again this reflects a youthful age structure, as well as higher incomes and better healthcare in urban areas.

Table 2.8 Natural increase and urban growth per annum in Bangladesh and India in 2009

	CBR/1,000		CDR/1,000		Natural increase (% per annum)		Urban population growth (% per annum)
	Rural	Urban	Rural	Urban	Rural	Urban	
Bangladesh	21.7	17.5	6.0	4.4	1.57	1.31	2.81
India	25.2	18.8	8.1	6.0	1.70	1.28	2.28

Rural–urban migration

Natural population increase in towns and cities explains only part of the contemporary urbanisation in the developing world. Table 2.8 shows that natural increase in rural Bangladesh and India in 2009 outstripped natural growth in towns and cities. However, in Bangladesh less than half of all urban growth was due to natural increase in urban areas. In India it accounted for 44% of urban population growth. The remaining urban population increase is therefore due to internal rural–urban migration. This means that urbanisation in Bangladesh is a function of natural increase *and* rural–urban migration. This situation is typical of many LEDCs, not just in south Asia, but throughout the developing world.

The causes of rural–urban migration

Urbanisation in the developing world is driven by two sets of factors.

● The concentration of investment and employment in towns and cities. Productive industries and services cluster in cities: 80% of the world's GDP is generated by urban areas.

● Poverty, inequality, and lack of investment and economic opportunity in rural areas.

In Bolivia migration from rural to urban areas is triggered by poverty and lack of opportunity in the countryside and the perceived economic and social advantages (employment, income, education) of cities. Urban wages, for example, are on average four times those in the countryside. Around two-thirds of the rural population live in extreme poverty, compared with approximately one-quarter in urban areas. In recent decades, rural poverty has increased because of drought and the collapse of the tin mining industry.

Just over half of all internal migration is directly from the countryside to urban areas. However, 27% is from smaller towns to large cities, especially to the capital La Paz/El Alto. This suggests that many migrations occur as a series of steps: first from the countryside to the nearest town, and later from small towns to major regional centres and the capital.

Now test yourself

26 What is the difference between million cities and mega cities?

Answer on p. 123

Tested

Typical mistake

The rapid growth of million cities and mega cities is not just fuelled by rural–urban migration. Migration from small urban centres to larger ones (urban–urban migration) is a feature of many countries in the developing world.

Examiner's tip

Contemporary urbanisation is the result of rural–urban migration *and* natural urban population growth in LEDCs. The contribution of each varies between countries. Detailed understanding of the relative importance of each factor is compromised by lack of data on urban and rural growth rates in most LEDCs.

Now test yourself

27 Describe three push factors which generate rural-urban migration in the developing world.

Answer on p. 123

Tested

Case study **Lagos: city of slums**

Urban growth. Lagos, on the coast of Nigeria, is Africa's second largest city and one of the fastest growing in the world. From a population of just 300,000 in 1950, by 2010 Lagos had become a sprawling mega city of more than 10 million people (Table 2.9). Currently it is on track to reach 19 million by 2025.

Three-quarters of the city's population growth is due to internal rural–urban migration and international migration from neighbouring Benin and Togo. Rural–urban migration emphasises the extreme poverty of rural areas with migrants flooding into the city at a rate of 10,000 per week.

Table 2.9 Lagos: actual and projected population growth, 1995–2025

	Population (millions)	**World ranking by size**
1995	5.84	29
2000	7.28	28
2005	8.86	24
2010	10.79	20
2015	13.12	14
2020	15.83	12
2025	18.86	11

Slums: the impact of urbanisation. Rapid urbanisation in Africa has not been accompanied by comparable rates of economic growth and industrialisation. The result is mega cities such as Lagos, described as 'one of the largest concentrations of urban poverty on the planet'. Lagos is unplanned, chaotic and bursting at the seams.

The city lacks the financial resources to provide its population with basic needs and services such as shelter, sanitation, rubbish collection, clean water supply, etc.

Despite recent economic growth Lagos cannot keep pace with the demand for jobs. Thus 70% of Lagos's population live in poverty. Like other cities in the developing world, the economy relies heavily on the informal sector to absorb surplus labour.

Self-built slums reflect the inability of the city to provide even the most basic housing and the scarcity of building land. There are more than 100 slum districts in Lagos, and seven out of ten people are slum dwellers. Some slums, such as Ajegunle with a population of 1.5 million, are city-sized. Immigrants squat illegally on any parcels of undeveloped land. Much of the land is swampy and threatened with flooding during the rainy season and at high tides. Slum dwellers also live in constant fear of eviction.

The Makoko slum accommodates 300,000 people in shacks built on stilts in a tidal lagoon. There is no sanitation and waste and raw sewage are discharged into the lagoon. Slum population densities often exceed 100,000 persons/km². Three-quarters of slum dwellers live in single-room households, which have an average of 4.6 persons. Without planning and legal title the slum settlements lack essential services such clean water supply and sanitation.

Lagos's exploding population, appalling poverty and chronic unemployment are also responsible for high levels of crime such as protection rackets and drug dealing. Meanwhile, corruption, especially among the city's elite, is endemic.

Examiner's tip

In LEDCs slums are referred to by general names such as *squatter settlements*, *informal settlements*, and *shanty towns*. The term *squatter settlement* should only be used when residents have no legal title to the land they occupy. However, all self-built slums are ad hoc constructions and are accurately described as *informal*.

Typical mistake

Urbanisation in the developing world is not the same as that experienced by MEDCs in the nineteenth and twentieth centuries: it is occurring faster and against a background of more rapid population growth. In Africa it is also taking place without significant economic growth and industrialisation.

Case study **London and the cycle of urbanisation**

Patterns of urban population change in MEDCs appear to have come full circle over the past two centuries. This change is referred to as the **cycle of urbanisation**. It consists of four stages:

- rapid urbanisation: associated with nineteenth-century industrialisation and the concentration of population in the inner zones of towns and cities
- suburbanisation: the decentralisation and internal redistribution of population from inner to outer zones
- counterurbanisation: continuing decentralisation but with the dispersal of urban populations to smaller towns and rural areas
- re-urbanisation: the return of population to live in the inner zones of cities

London, a mega city (12.5 million people live in the metropolitan area), has experienced the cycle of urbanisation over the past 200 years.

Urbanisation. In the early nineteenth century London was by far the biggest city in Europe, with an estimated population of

1 million (Figure 2.15). Britain's industrial revolution and the status of London, not just as the capital, but as the organising centre of a vast empire and the country's leading port, led to explosive population growth in the nineteenth century. By 1900, London's population had reached 6.3 million. This spectacular growth was mainly due to in-migration.

For most of the nineteenth century London's population growth was concentrated in the inner zone, within a couple of miles of Trafalgar Square. Densities were high, living conditions squalid and insanitary, and death rates often exceeded birth rates. The inner London parish of Tottenham was typical: its population grew from 3,600 in 1801 to 137,000 in 1901. During this time its average population density increased from 193 to 7,280 persons/km².

Suburbanisation. Transport innovations such as the railways, electric trams and motorised buses were the catalysts for early population decentralisation from 1850 to the 1930s. Thanks to the railways, by the 1850s and 1860s the first commuters were already moving to locations outside London. Villages such as Bromley and Romford became the first **exurbs**.

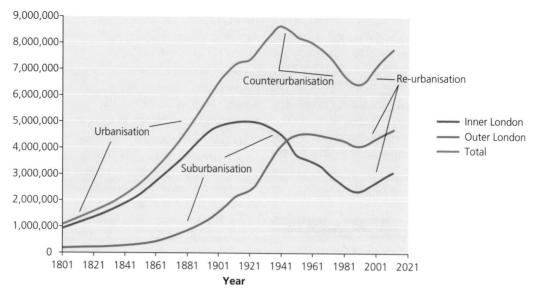

Figure 2.15 Population change in London, 1801–2011

The growth of the suburbs really took off with the development of intra-urban transport. First, horse-drawn trams and buses, and then in the 1890s and 1900s electric trams and motorised buses. Urban growth extended radially following the new tram and rail routes. Developers, builders and railway companies cooperated to promote suburban development. For example, in the 1920s the Metropolitan Railway (also a property developer) marketed 'Metroland' — northwest London including parts of Middlesex, Hertfordshire and Buckinghamshire — to encourage house building for middle-class commuters and expand its passenger traffic.

Following the nationalisation of transport in London in 1933, the Metropolitan Railway was absorbed into the London underground. The major boost to suburbanisation occurred in the 1920s and 1930s when the underground network was extended beyond the northern suburbs to incorporate places such as Edgware, Cockfosters and Stanmore. A massive programme of house building followed. Soon the suburbs became an endless sprawl of low density semi-detached and detached housing.

Manufacturing industry, geared to road transport, also shifted to the suburbs in the inter-war years, with light industries, located in modern industrial estates, developing along radial routes such as the Great West Road.

With only limited urban development beyond the suburbs, Greater London's population peaked in the 1930s at 8.6 million.

Counterurbanisation. Decentralisation resumed, but on a more expansive geographical scale in the 1960s and 1970s, as middle to higher income groups began to move out beyond the suburbs, targeting smaller towns and villages (exurbs) beyond the green belt but within commuting distance of the capital. The main attractions were lower housing costs, a pleasant environment, and less pollution, crime and congestion. Improvements to the road network (motorways) and faster commuter trains assisted counterurbanisation. By the late 1980s London had lost one-quarter of its inhabitants, though London's wealthiest inhabitants remained in elite residential districts in the central zone such as Kensington and Chelsea.

Post-war planning and the successful new town programme also drove decentralisation. New towns such as Crawley and Stevenage were built with public funds to solve London's housing shortages and provide inner London families with better standards of housing. Originally eight new towns were planned. Later a second wave of new towns, based on established urban centres such as Northampton and Peterborough, was designated.

Re-urbanisation. In the past 30 years a return to living in central London has become fashionable. This has led to a population revival known as re-urbanisation. Decentralisation had made journeys to work longer, with commuter times extended by chronic congestion on London's roads. Deregulation of the City in 1986 gave impetus to London's position as a leading centre for global financial services. Rising prosperity among City workers made accommodation in central London affordable. Proximity to City offices and easy access to shops and leisure services added to the attractions of city-centre living — a lifestyle that appealed to young single people and couples with higher incomes. Employment in the City grew by 50,000 between 1991 and 2001 to more than 300,000. Many were skilled foreign workers.

Initially re-urbanisation was associated with the **gentrification** of former low-income neighbourhoods such as Islington and Notting Hill. Terrace houses built in Victorian and Edwardian times were upgraded through private investment, driving up house prices and creating enclaves dominated by professional groups. Today, gentrification is extending into previously unfashionable areas, such as Tooting and Brixton.

Public investment also boosted re-urbanisation. London's derelict docklands were regenerated by public and private capital between 1981 and the early 1990s. Between 1981 and 1998, 25,000 new homes were constructed, many of them upmarket apartments with waterfront locations east of the City.

London's population decline in the 1970s and 1980s has been reversed in the past 20 years. In the decade 2001–2011 the capital's population grew by 12% — faster than any other region in England and Wales. London currently has 9 out of the 20 fastest-growing authorities in the UK, and Newham and Tower Hamlets recorded an intercensal population growth 2001–2011 in excess of 20%. This rapid growth is due to international immigration. The urbanisation cycle has come full circle with renewed urbanisation — this time triggered by international rather than internal migration.

Now test yourself   Tested ☐

28 Define the following terms: urbanisation, suburbanisation, counterurbanisation and reurbanisation.

29 Draw a diagram, with explanatory notes, to show the main features of the cycle of urbanisation.

Answers on p. 123

Mega cities and sustainable growth   Revised ☐

The consequences of rapid urban growth

It is likely that the world's population will increase from 7 billion to 9.3 billion between 2011 and 2050. Towns and cities will absorb all of this increase, and 95% of growth will take place in the developing world. The biggest urban expansion will be in Asia (an extra 1.4 billion) and Africa (0.9 billion). Uncontrolled urban growth in mega cities and million cities, detached from significant economic and social development, threatens to create a nightmare scenario of extreme poverty, slums, resource depletion and environmental degradation — a process already underway in Sub-Saharan Africa.

Sustainable urban living

The concept of sustainable cities has particular urgency as a combination of rapid urbanisation and rising material aspirations by billions of people in the developing world begins to have calamitous effects on the environment.

The solution is sustainable growth: growth that does not degrade the environment and that uses the planet's resources — fresh air, clean water and energy — at rates that equal their renewal by natural processes. However, sustainability will require reductions in consumption. And given that 1 billion people in the developing world already live in urban slums, cutting consumption seems an unrealistic hope.

For the foreseeable future, cities are likely to remain (a) net consumers of land, energy, water, food and other resources, and (b) dependent on natural systems to absorb their waste products. In practice, sustainability will mean slowing rates of development and living within **ecological limits**. What is certain is that unless we take action to slow development or live within ecological limits, the consequences for the environment and for future generations are bleak.

Ecological footprints

An ecological footprint is 'the amount of productive land and water a given population requires to support the resources it consumes and the absorption of its wastes'. The smaller the **ecological footprint** the closer a city or society comes to sustainability.

Ecological footprints are measured in units of global hectares (gha). One gha has an annual biological productivity equal to the world average. The global biosphere has a total of 11.2 billion gha, which averages out at 1.8 gha for every person on the planet. However, actual usage is 2.2 gha per person, indicating an annual drawdown of natural capital. London, for example, needs 49 million gha to produce and dispose of all the energy and materials needed to support its 8.1 million inhabitants. This averages out at 6.63 gha per person, which is three times the global average.

Despite significant efforts in the past decade to reduce carbon emissions, recycle waste, conserve water supplies, reduce energy consumption and so on, ecological footprints for UK cities average 3.1 gha — nearly twice that needed for sustainability.

China's legacy of unsustainable urban growth

In the past three decades China has experienced massive industrialisation and urbanisation. The year 1979 was a watershed in China's economic and political history. The communist government relaxed its rigid economic policies and moved towards a more open, market-based economy. At this time 80% of China's population lived in the countryside. Within 30 years, industrialisation had dramatically changed China's population distribution. By 2009 45% were urban dwellers — a proportion that will rise to 60% by 2030. On current trends, China's urban population will reach 926 million in 2025, and by 2030 pass the 1 billion mark.

Future urban expansion on this scale will present China with huge environmental challenges. Recent urban growth has put acute pressure on land, energy, water, wildlife and other natural resources. Urban air pollution is a major concern. In 2007 the World Bank reported that 16 of the world's 20 most polluted cities were in China. Severe air pollution results from hundreds of coal-fired power plants and large concentrations of heavy industries such as steel making and chemicals. Particulate and sulphur dioxide emissions cause smog, acid rain and dry deposition as well as respiratory illness, heart disease and lung cancer.

Of equal concern is the soaring number of motor vehicles. Beijing already has 3.6 million motor vehicles and by 2030 China will have more motor vehicles than the USA. Motor vehicles emit nitrogen oxide and ozone (both injurious to human health), and create **photo-chemical smogs** that hang over major cities in the summer months. Meanwhile, pollution of rivers and lakes by industrial effluent, agricultural runoff and untreated sewage adds to environmental pollution loads, contaminates drinking water and damages health.

> **Photo-chemical smog** occurs when sunlight interacts with oxides of nitrogen and hydrocarbons emitted from vehicle exhausts. The result is high concentrations of pollutants such as ozone. The smog is a feature of large cities in sunny climates, e.g. Los Angeles, Beijing.

Planning for sustainable cities: ecopolis

China leads the way in planning the world's first fully sustainable eco-cities. China's planned eco-cities have a number of features:

- they aim to be **carbon neutral** and rely on renewable energy such as wind and solar power
- they will collect and recycle rainwater and use green building technologies such as grass roofs
- the built environment will include extensive green spaces to promote evaporation and cooling in summer
- houses will be well insulated and energy efficient and residents will be able to walk or cycle rather than make journeys by car
- public transport will replace private car ownership

Dongtan, on the alluvial island of Chongming at the mouth of the Yangtze River, provides the model for future Chinese eco-cities. Dongtan occupies an 86 km² site and has a target population of 500,000. It aims to:

- be carbon-neutral with an ecological footprint only a fraction of those of Beijing (3.86) and Shanghai (3.42)
- produce zero waste
- generate its own energy from clean renewable resources such as wind power, photovoltaic cells and biofuels

Exam practice answers and quick quizzes at **www.hodderplus.co.uk/myrevisionnotes**

- incorporate green building technologies to reduce the demand for energy to heat and cool homes
- provide its own water supply by collecting rainwater, and storing it in Dongtan's canals and other water features
- be a low-rise and low-density urban environment (by Chinese standards), with houses separated by forests and organic farms, lakes and tourist attractions
- have a compact layout, and be largely car-free; residents will cycle or walk to get to work, shops, services and schools

Each Dongton resident will have on average 27 m² of green space, which compares favourably with cities such as London and Los Angeles. Vital infrastructure, including the bridges and tunnel linking Chongming Island with downtown Shanghai, have already been completed.

However, despite initial optimism, progress on the Dongtan ecopolis has been slow. The original plan to accommodate 50,000 residents by 2010 has now been shelved and the entire project is on hold. Indeed, it is even possible that the scheme could be abandoned altogether and the site sold. Meanwhile, a new road link brings Chongming Island within commuting distance of Shanghai and threatens to replace the radical eco-city with a conventional green suburb for Shanghai's middle classes. This would mean more urban sprawl and more loss of farmland and wildlife habitats.

Typical mistake

Remember that there is a distinction between cities that aim to reduce or minimise their impact on the environment (e.g. London) and those such as Dongtan that plan for full sustainability. Progress towards sustainability for the foreseeable future will result in only a small reduction in the ecological footprints of cities.

Global challenges for the future

Globalisation: a two-speed world   Revised

Globalisation has both positive and negative impacts. These impacts are economic, social, demographic and environmental.

Poverty

Globalisation has increased the sum total of wealth, but its impact varies widely between countries and income groups. On the positive side, during the period of rapid globalisation since 1980, the incidence of extreme poverty has fallen in all major regions of the developing world (Table 2.10).

Enquiry question: What are the social and environmental consequences of globalisation and can we manage these changes for a better world?

Table 2.10 Poverty in the developing world: per cent of population living on less than US$1.25/day, 1981–2008 (Source: World Bank)

	1981	2008
East Asia and the Pacific	77	14
South Asia	61	3.6
Middle East and North Africa	14	6.5
Latin America and Caribbean	16.5	2.7
Sub-Saharan Africa	51	47

In 1981, 43% of the population in the developing world lived in poverty. By 2010, this proportion had halved. Even so, that still leaves around 1 billion people suffering poverty; and most people who have moved above the poverty line are still poor by the standards of middle- and high-income countries.

China has achieved the most spectacular success in alleviating poverty. Liberalisation of China's economy in 1979 triggered rapid industrialisation and sustained economic growth. As a result 500 million Chinese were removed from poverty between 1981 and 2004.

Typical mistake

Although the proportion of people suffering poverty has fallen everywhere, in Sub-Saharan Africa (SSA) there has been an absolute increase. Rapid population growth in SSA has pushed an extra 180 million people into poverty.

Examiner's tip

Progress in fighting poverty has been most impressive in countries where (a) recent economic development has occurred on a significant scale and (b) population growth has been contained within reasonable limits.

Inequality and the poverty gap

We live on an unequal planet where income inequalities have increased in most low- and middle-income countries.

- The wealthiest one-fifth of humankind takes four-fifths of global income.
- The share of the poorest 40% has increased by only 1% since 1990.

Globalisation has failed to narrow the gap between rich and poor. Inequality has increased most dramatically in China. Chinese urban dwellers enjoy an annual average disposable income per capita that is three times greater than their rural cousins. And this rural–urban divide is getting larger.

Globalisation has also increased income disparities in the developed world. Salaries of bankers and other City workers in London's financial and commercial services have increased massively in the past 30 years, creating a super-rich elite. Meanwhile, those excluded from the global economy (e.g. most residents in neighbouring Tower Hamlets and Hackney) have seen little improvement in their incomes.

Typical mistake

It is not surprising that globalisation creates inequalities. Globalisation is an expression of capitalism, and as such this inevitably producers 'winners' and 'losers'. The hope is that new wealth created will eventually 'trickle down' and benefit the poor.

Examiner's tip

You should recognise that FDI has economic advantages and disadvantages for all countries.

Foreign direct investment

Globalisation has dispersed some economic activity from core industrial regions such as the USA, EU and east Asia to low- and middle-income countries. This process of **offshoring** lowers costs and gives TNCs direct access to foreign markets. Countries benefit from FDI flows, gaining much needed employment, exports, and technological and commercial expertise. The downside to offshoring is the loss of jobs in the developed world.

However, FDI has a very uneven geographical distribution. Until recently the main flows of FDI were between North America, Europe and east Asia. This pattern is beginning to change. In 2010, for the first time, more than half of global FDI went to developing and transitional economies such as Brazil, Russia, India, China and South Africa (BRICS). Many large TNCs (e.g. Coca Cola, Toyota, GlaxoSmithKline) depend heavily on investments in the BRICS and other transitional economies (Figure 2.16).

Even so, FDI has largely by-passed the world's least developed countries (LDCs). Although the 48 LDCs accounted for 12% of the world's population in 2010, their share of global FDI was just 2%. And most of this is concentrated in small enclaves specialising in the export of primary products such as oil and mineral ores.

International trade

International trade has expanded threefold since 1990. Geographically the pattern of international trade is very uneven, being dominated by wealthy countries in Europe and North America, and emerging economies in Asia (particularly China and India) (Figure 2.17). In contrast, Africa has barely 3% of world trade in merchandise and 2% in services. This underlines the limited participation of the world's LDCs in the global economy. It also suggests that the expansion of world trade has brought them few benefits.

The World Trade Organisation (WTO) is the international body that deals with the rules of trade at global level. It promotes free trade by removing obstacles such as tariffs and subsidies and settles trade disputes between member states. The WTO is dedicated to **trade liberalisation**. Trade liberalisation expands the volume of trade, and should, ultimately, benefit

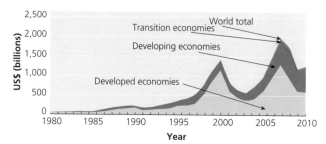

Figure 2.16 FDI flows, 1980–2010 (Source: UNCTAD)

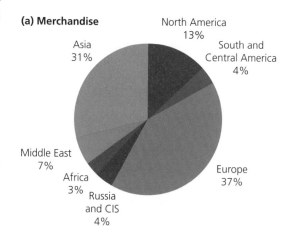

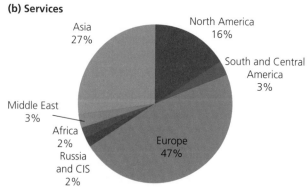

Figure 2.17 Distribution of world trade in merchandise and services, 2011

everyone. However, trade liberalisation also has undesirable side effects. LDCs and other low-income countries may be unable to withstand competition. In these circumstances it is prudent to maintain protective trade barriers.

The moral and social consequences of globalisation

Revised

Many observers argue that globalisation has unwanted moral and social consequences; that driven by profit, business enterprises seek to minimise their costs by exploiting workers in low-income countries. Meanwhile TNCs based in the developed world expand their markets globally, promoting consumerism with often adverse effects on local cultures and the health of local inhabitants.

Globalisation and the exploitation of workers

Working conditions that would not be tolerated in the western world ('sweat shops') flourish in the developing world because (a) people are desperate for work and (b) employment legislation is weak and often not enforced.

Fila, a South Korean TNC, is a leading brand in sports shoes and other sportswear. Like other leading sportswear companies, it does not manufacture. Because sportswear manufacture is labour-intensive, production is offshored (and subcontracted) to countries with low-wage economies in Asia, Latin America and eastern Europe.

However, the drive to reduce costs is often at the expense of wages and working conditions for employees. Fila, like other TNCs, push their suppliers for shorter delivery times and lower unit costs. If these demands are not met contracts may be cancelled and production workers may face redundancy.

Globalisation and local culture

Globalisation and the uniformity of culture

Globalisation includes the worldwide diffusion of western culture. In practice this has meant the diffusion of US culture or **Americanisation**. Americanisation promotes **consumerism**, exploiting people's material desires. Described as a kind of 'cultural imperialism', Americanisation often weakens local cultures. The result is bland uniformity as people abandon their traditions and adopt western tastes and lifestyles.

This process, exemplified by the US fast food chain McDonald's, has become known as **McDonaldisation**. Fast food restaurants such as McDonald's, Burger King and Subway have developed a highly successful marketing formula, selling standardised, predictable products cheaply and efficiently. Their success (McDonald's and Subway each have more than 30,000 restaurants worldwide) threatens traditional foods and diets, both important aspects of local culture.

Americanisation is driven by powerful advertising campaigns, as well as by western films, television programmes, sports sponsorship, popular music, etc.

Globalisation and cultural diversity

Globalisation can foster cultural diversity as well as uniformity. McDonald's in India sells products such as McAloo Tikki burgers and and Chicken Kebab burgers designed for the local market. A reformulated Lucozade, with a more intense flavour, is marketed in China. This process, whereby foreign TNCs modify products to sell in local markets, or source commodities or parts locally, is called **glocalisation**.

International migration stimulated by globalisation has created cultural diversity in many MEDCs. In London in 2011, more than half the populations of Brent, Chelsea and Kensington, Newham and Westminster were foreign born. Multiculturalism has allowed cultural diversity to flourish. Nowhere is this more apparent than in the high street in the UK, with shops such as *Polski Sklepp*, Indian curry houses, sushi bars, Thai restaurants and Chinese takeaways. Other expressions of cultural diversity include ethnic popular music such as rap, Bollywood films, and events such as Afro-Caribbean carnivals and Asian cultural festivals or *melas*.

Globalisation and human health

The expansion of TNCs into the developing world, manufacturing and marketing pharmaceutical products and consumables such as tobacco, fast food and soft drinks, can have a significant impact on human health.

TNCs and the sale of life-saving drugs

Diseases such as malaria, tuberculosis and HIV/AIDS cause massive suffering and premature death throughout the developing world. Malaria, endemic in 109 countries, killed 863,000 people in 2009. Yet, life expectancy and morbidity could be improved for millions of people if they had access to affordable drugs. Pharmaceutical TNCs based in MEDCs, however, control their manufacture and sales. For TNCs the priority is profit and recouping investment in research and development. As a result many life-saving drugs, patented by pharmaceutical companies, cannot be manufactured cheaply under licence in LEDCs. This makes them unaffordable to millions of people in the developing world. Furthermore, because profit margins for selling drugs in poorer countries are small, there is limited incentive for pharmaceutical TNCs to develop new drugs for this market.

TNCs and the globalisation of food systems

Many TNCs manufacture and market food and other consumables knowing they are injurious to human health. Examples include tobacco, fast food and powdered milk.

● Worldwide approximately 1.3 billion people smoke cigarettes or other tobacco products. In MEDCs the health risks from smoking are well known, therefore the market for tobacco products in these countries has dwindled. Tobacco companies have responded by targeting LEDCs. Today more than 80% of smokers are in LEDCs and the numbers are increasing rapidly.

● Fast food chains have successfully promoted foods high in sugar and saturated fats in LEDCs. Supported by intensive advertising, fast food is now a globalised industry and is expanding rapidly into markets in the developing world. The adverse effects on human health include obesity and increased risks of diet-related diseases.

● TNCs such as Nestlé have promoted the sale of powdered milk products for infants as a replacement for breast feeding. Breast feeding has physical benefits both for babies and mothers. Yet in the Philippines, food companies spend US$100 million a year advertising breast milk substitutes. Reliance on powdered milk increases the risk of malnutrition in infants and increases pressure on the budgets of families who are already poor.

Reducing the environmental and social costs of globalisation

Revised

Managing the environmental and social consequences of globalisation requires action at a variety of scales, from local to global.

Local actions

Waste management: landfill

The disposal of solid waste generated by towns and cities presents major environmental problems. London generates 4.4 million tonnes of municipal household waste a year. Most of this is disposed in 18 **landfill** sites — some

Now test yourself

30 How has globalisation affected local cultures?

Answer on p. 123

Tested

Examiner's tip

A balanced approach is needed to any assessment of the impact of TNCs on global health. Any adverse effects must be considered against the background of investment and employment created by TNCs.

Now test yourself

31 State three ways in which globalisation has adversely affected human health.

Answer on p. 123

Tested

more than 120 km from the capital. Landfill is an environmentally unsatisfactory method of waste disposal for three reasons:

- leakage of toxic chemicals into the environment
- emissions of GHGs such as methane and carbon dioxide
- loss of countryside and amenity as well as brownfield sites (e.g. old quarries) which could be used for other purposes in crowded urban regions

Now test yourself

32 What are (a) landfill sites, (b) brownfield sites?

Answer on p. 123

Tested

Recycling

The UK is rapidly running out of landfill space and needs to find alternatives to comply with EU directives (Figure 2.18). One approach is to extend **recycling** schemes and reduce the amount of waste going to landfill. Local authorities now require households to sort their recyclable domestic refuse. As a result the proportion of domestic waste recycled in England increased from 11% to 40% between 2001 and 2011. And in 2011 more commercial and industrial waste was recycled than was sent to landfill.

Further reductions in landfill will be needed in future. New technologies such as **anaerobic digestion** will help reduce organic waste. Some councils have introduced fortnightly rather than weekly kerbside bin collections, and a few have considered charging householders for the amount of waste they produce.

Recycling also benefits the environment, currently reducing carbon dioxide emissions in the UK by 18 million tonnes per year.

Anaerobic digestion is the breakdown of biodegradable material in the absence of oxygen by microbes to produce methane and carbon dioxide.

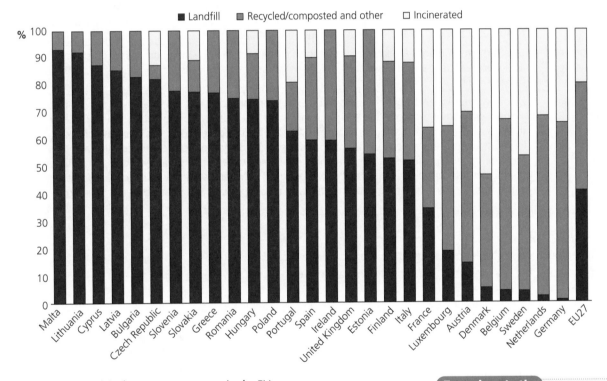

■ Landfill   ■ Recycled/composted and other   □ Incinerated

Figure 2.18 Municipal waste management in the EU

National actions

See Unit 1, 'Tackling the causes of climate change', 'National strategies' on page 34, i.e. subsidies for cleaner technologies, carbon taxes and expansion of renewable energy.

International actions

See Unit 1, 'Tackling the causes of climate change', 'International strategies' on page 33, i.e. Kyoto Protocol, cap and trade/carbon credits; and 'Meeting the challenge of global warming', 'Reducing carbon emissions', on page 39.

Examiner's tip

In your revision it is important to make connections with topics covered in Unit 1. See earlier in this book, particularly the topics 'Coping with climate change' (p. 33) and 'The challenge of global hazards for the future' (p. 37).

Strategies for conserving resources

Approaches to conserving and managing resources and creating a more equitable world include green strategies and ethical purchasing.

Green strategies

Green strategies are policies to manage human activities with a view to prevent or minimise potential harmful effects on the natural environment (Table 2.11).

> **Green strategies** are plans developed by governments and businesses that concern environmental and sustainability issues and promote, for example, recycling, waste reduction, energy and water saving, biodiversity, etc.

Table 2.11 Green strategies in the UK

Strategy	Viability
Recycling	Recycling of domestic and industrial waste is well established in the EU. Under an EU directive, all EU member states must compost or recycle half their household waste by 2020. This reduces pressure on mineral, biotic and energy resources and landfill sites. It also cuts carbon emissions from waste incinerators and methane from green waste.
Green taxes	In the UK green taxes, which aim to reduce emissions of carbon and other pollutants and protect the environment, include fuel duty, air passenger duty, road tax, landfill tax, etc. These taxes are unpopular with businesses because they increase costs and with households because they add to the cost of living. Green taxes are widely perceived as raising revenue for the Treasury, with the extra income not being spent on the environment.
Subsidies for renewable energy	UK government subsidies boost green energy and help meet international obligations to reduce carbon emissions and achieve the target to generate 20% of energy from renewables by 2020. Subsidies are available for wind, biomass, solar and tidal power. Subsidies should stimulate private investment in renewables, but uncertainty about the government's long-term commitment has deterred some investors. Wind power expansion arouses opposition from environmentalists and communities affected by wind farms. Domestic solar energy production is promoted by subsidies paid to households for the electricity they feed into the national grid.
Grants to improve household energy efficiency	Government funding is available to households to improve their heating and energy efficiency. Grants are income-related, and are available for loft insulation, cavity wall insulation, draught-proofing and central heating.
Modifying lifestyles	Modifying individual lifestyles reduces carbon emissions and conserves natural resources. Examples include: walking or cycling rather than motoring for short journeys, using public transport rather than cars and taxis, purchasing locally grown foodstuffs to reduce food miles, avoiding unnecessary long-distance travel, reducing consumption of meat and dairy products, collecting rainwater from roofs for irrigating gardens, composting biodegradable household and garden waste, turning down thermostats on central heating systems, switching off lights and stand-by functions on electrical appliances, recycling, avoiding consumables with excess packaging and using plastic bags from supermarkets.

Ethical purchasing

Ethical consumer behaviour involves purchasing goods that concern ethical issues such as fair trade, the environment, human rights, animal welfare and labour conditions.

Fair trade

Consumers may choose fair trade products such as chocolate, tea, coffee and bananas in the belief that their purchases improve the wages and working conditions of farmers and their families in the developing world. Cocoa farmers in West Africa, for example, receive only a tiny fraction of the price of chocolate bars sold in the developed world. Thus while the retail price of chocolate has increased by 60% in recent years, the price paid for cocoa beans has almost halved. The chocolate industry, dominated by large TNCs such as Nestlé, Mars and Kraft (Cadbury's), makes it difficult for small growers to get a fair price for their cocoa.

International trade is weighted in favour of rich countries and powerful TNCs. Most poor countries not only have limited bargaining power but also depend heavily on commodities whose price fluctuates wildly on world markets. Moreover the real value of these commodities tends to decrease over time compared with manufacturing goods and commercial services.

> **Now test yourself**
>
> 33 Give four examples of green strategies in the UK.
>
> **Answer on p. 123**
>
> Tested

Local sourcing

Ethical purchasing also applies to products sourced locally. Buying produce sold in farmers' markets, for example, helps to support local employment, reduces food miles and often pre-empts the need for environmentally unfriendly packaging. Many consumers, for health, animal welfare and environmental reasons, are also prepared to pay premium prices for organically grown food.

Fair trade and ethical purchasing make only small contributions to conserving environmental resources and creating a more equitable world. The priority for most buyers is price, and ethical products invariably are more expensive than commercial ones.

Examiner's tip

When revising this topic it is important to (a) evaluate the effectiveness of green strategies and ethical consumption and (b) clarify your own view on these initiatives.

Now test yourself

Tested ☐

34 What are the economic and environmental advantages of the local sourcing of food?

Answer on p. 123

Examiner's summary

World cities

- ✔ It is important to have a clear understanding of the differences between urbanisation, counterurbanisation, suburbanisation and re-urbanisation.
- ✔ The population movements that drive the growth of mega cities are not just rural–urban migrations; also important is urban–urban migration, from smaller towns and cities to mega cities.
- ✔ Urbanisation in the developing world is due to a combination of migration and high rates of natural increase within urban areas. Owing to a lack of data, it is difficult to assess the relative importance of each factor.
- ✔ Urbanisation in the developing world is often not underpinned by a commensurate rate of economic growth.
- ✔ The cycle of urbanisation comprises population movements which are: inter-regional (urbanisation, counterurbanisation) and intra-urban (suburbanisation, re-urbanisation); and centripetal (urbanisation, re-urbanisation) and centrifugal (suburbanisation, counterurbanisation).
- ✔ It is debatable whether urban growth and cities can ever be truly sustainable.

Global challenges for the future

- ✔ Trade liberalisation increases global wealth, but does not guarantee that this wealth is shared equally between rich and poor countries.
- ✔ Progress in reducing poverty in the developing world has been most successful in countries which have (a) enjoyed rapid economic growth, and (b) contained their population growth within reasonable limits.
- ✔ Globalisation, driven by capitalism, inevitably results in inequalities; the hope is that the new wealth it creates will 'trickle down' and eventually benefit everyone.
- ✔ In preparation for discussion, clarify your personal viewpoint on issues connected to the moral and social consequences of globalisation.
- ✔ Issues that surround TNCs and the economic, social and environmental impact of globalisation are complex. It should be realised that the activities of TNCs in the developing world have benefits as well as disbenefits for local people and environments.

Exam practice

Section A

1 Study Table 2.12.

Table 2.12

	Index of globalisation 100 = maximum	G8	OPEC
Belgium	92.76		
Austria	90.55		
Switzerland	86.64		
Norway	83.19		
Malaysia	77.43		
USA	74.88		
Nigeria	58.01		
Paraguay	57.53		
Congo DR	36.60		
Afghanistan	35.35		

    (a) Which country in Table 2.12 belongs to (i) the G8 and (ii) OPEC? Put ticks in the appropriate boxes. [1]

    (b) What evidence in Table 2.12 connects globalisation to economic development? [2]

    (c) What are LDCs? [3]

    (d) Explain how trade blocs such as the EU help member countries to increase their wealth and power. [4]

2 Study Figure 2.6 on p. 58.

    (a) Describe the changes in the infant mortality rate in the UK between 1900 and 2010. [3]

    (b) Suggest reasons for the changes you described in question 2(a). [4]

    (c) Explain the factors responsible for the variations in the fertility rate in the UK since 1900. [4]

3 Study Figure 2.16 on p. 76.

    (a) How did the developing world's share of FDI change between 1980 and 2010? [2]

    (b) Outline the processes responsible for the changes described in question 3(a). [3]

    (c) Explain how globalisation has stimulated the growth of mega cities in the developing world. [3]

    (d) What evidence suggests that new urbanisation in the developing world is unsustainable? [4]

Section B

4 Study Figure 2.11 on p. 64.

    (a) Suggest why such a large volume of international migration focuses on Europe and North America. [10]

    (b) Examine the role of communications and transport technology in accelerating the influence of globalisation. [15]

5 Study Figure 2.17 on p. 76.

    (a) Suggest reasons for the unequal distribution of world trade in merchandise and services. [10]

    (b) Explain how TNCs play a crucial role in the spread of global business and trade. [15]

Answers and quick quiz 2 online

Online

Exam practice answers and quick quizzes at **www.hodderplus.co.uk/myrevisionnotes**

3 Extreme weather

Extreme weather watch

Extreme weather phenomena Revised

Extreme weather phenomena are exceptional, high-intensity events that occur infrequently. They include storms, droughts, heat waves, severe spells of cold weather and heavy rainfall and snowfall (Table 3.1).

> **Enquiry question**: What are extreme weather conditions and how and why do they lead to extreme weather events?

Table 3.1 Nature and distribution of extreme weather phenomena

	Nature	Distribution
Tropical cyclones	Large, violent revolving storms associated with hurricane-force winds and torrential rain.	The tropics and sub-tropics — mainly in the North Atlantic and North Pacific Oceans, and the western Pacific.
Temperate storms	Large, mobile low-pressure systems that bring heavy rain and gale-force winds.	Middle–high latitudes, between 35° and 70°, e.g. NW Europe.
Tornadoes	Violently rotating, funnel-shaped columns of air in contact with the ground.	Continental interiors in mid-latitudes, e.g. Mid-West USA between the Rockies and Gulf of Mexico.
River floods	Rivers overtopping their banks because of excessive rainfall or rapid snowmelt.	Widespread. Flooding is most common in climates with wet seasons (e.g. monsoon) and where rivers drain high hills and mountains.
Blizzards	Heavy snowfall driven by strong winds.	High latitudes and mountainous regions.
Severe winter weather	Prolonged spells of abnormally low temperature and/or snow cover.	Middle–high latitudes. Continental rather than oceanic locations.
Heat waves	Prolonged spells of abnormally high temperature (e.g. >30°C).	Middle latitudes in summer.
Wildfires	Severe forest and scrub fires devastating extensive areas.	Regions with prolonged dry seasons (e.g. Mediterranean, California) and dense, combustible vegetation (e.g. conifer trees).
Droughts	Prolonged spells of abnormally low rainfall.	Widespread but most common in the sub-tropics (e.g. the Sahel), continental interiors (e.g. Mid-West USA) and the Mediterranean.

Now test yourself Tested

1 Name four types of extreme weather phenomena.

Answer on p. 123

Fieldwork and research Revised

Fieldwork data

Fieldwork, as a source of **primary data** related to extreme weather, consists of observations and recordings of:

- maximum and minimum daily temperatures
- rainfall, recorded at 24-hour intervals
- wind speed and wind direction, recorded at hourly intervals
- pressure, recorded at hourly intervals
- cloud cover and cloud types, recorded at hourly intervals

> **Typical mistake**
>
> Primary data are not just data collected through fieldwork observation and measurement. Unprocessed documentary data such as temperature, rainfall recordings, etc., available from the Met Office or newspapers, can also be classed as primary data.

Observation and recording may be made manually using maximum and minimum thermometers, a rain gauge, an anemometer and a barometer, or automatically using an electronic weather station linked directly to a personal computer.

Secondary data

Secondary data on weather phenomena are freely available from websites listed in Table 3.2. Access the data available on each website and consider how it could be used to investigate extreme weather events.

Table 3.2 Secondary data on weather phenomena

Website	
www.bbc.co.uk/weather	Five-day synoptic charts of the North Atlantic region.
www.weatheronline.co.uk	Current and 16-day forecast North Atlantic synoptic charts.
www.metoffice.gov.uk/public/weather/observations	Rainfall radar maps, updated every 30 minutes, showing rainfall distribution and intensity in the UK.
www.sat.dundee.ac.uk/	Weather satellite images of the UK, Europe and the world in infra-red and visible wavelengths.
www.metoffice.gov.uk	Hourly weather observations from local weather stations.
weather.noaa.gov/pub/fax/PTUK21.TIF	Sea surface temperature maps for the British Isles.
www.metcheck.com/ATLANTIC/jetstream.asp	Position of the Atlantic jet stream and forecast position over the next 7 days.
expert.weatheronline.co.uk/daten/profi/en/temps/temps.html	Tephigrams for six locations in the UK, showing vertical temperature distribution of the atmosphere at a given time.

Weather patterns and weather observations

Weather data from primary and secondary sources can be used to interpret weather patterns, such as air masses, depressions and anticyclones (Figure 3.1).

> **Examiner's tip**
>
> Websites such as those listed in Table 3.2 should be studied as part of your revision programme.

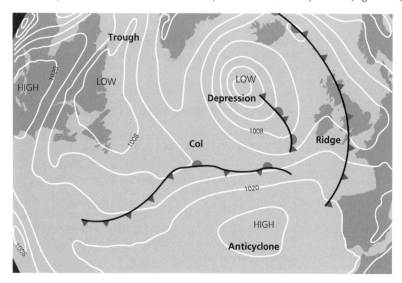

Figure 3.1 Common pressure patterns on synoptic charts

Air masses

Investigating air masses from fieldwork and secondary data might include:

- identifying the type of air mass affecting the UK or the region of observation
- deciding whether the air mass is **stable** or **unstable**
- observing and recording information on temperature, cloud cover, cloud type and precipitation over several hours
- comparing your own weather observations and records with what might be expected for a particular air mass at a given time of year

The passage of a depression

A predictable sequence of weather changes is associated with the passage of a depression. These changes, caused by the movement of weather fronts, affect temperature, pressure, wind direction, cloud and precipitation. They should be measured and recorded at the approach and passage of the warm and cold fronts either through direct observation (at 15- or 30-minute intervals), or through hourly data published online by local Met Office weather stations. The precise location of fronts and rain bands can be obtained from rainfall radar maps.

> An air mass is **stable** when the air temperature near the ground is lower than the air temperature above. Thus air near the ground displaced vertically will return to its original position (i.e. it is cooler and denser than the air above).
>
> An air mass is **unstable** when the air temperature near the ground is higher than the air temperature above. Air near the ground displaced vertically will not return to its original position (i.e. it is warmer and less dense than the air above).

Exam practice answers and quick quizzes at **www.hodderplus.co.uk/myrevisionnotes**

Anticyclonic conditions

Anticyclones affect the UK's weather and climate on approximately one day in four. Unlike depressions, they are often slow moving, and bring prolonged spells of calm, dry weather. Weather observations might include:

- the general synoptic situation shown on weather charts
- scrutiny of tephigrams from the nearest local weather station to (a) establish the existence (or otherwise) of a **temperature inversion** in the lower atmosphere and (b) confirm that the atmosphere is stable
- monitoring of cloud types and cloud cover. Some anticyclones bring clear skies, others overcast conditions and **stratus** (layer) clouds
- measurements of temperature (direct or from local weather stations at hourly intervals); extreme temperatures are possible under clear skies
- pressure, recorded at hourly intervals by local weather stations, to establish the barometric tendency (rising, falling, steady)

> A **temperature inversion** is an increase in temperature with height. This is the reverse of the normal vertical temperature trend — a lapse rate — in the lower atmosphere.

Examiner's tip

One possible approach to personal investigations into extreme weather is to compare (a) the standard model of a weather phenomenon with the data recorded in a specific event, or (b) the recorded data with the long-term averages for the period studied.

Now test yourself   Tested ☐

2 State five differences between depressions and anticyclones.
3 Describe the extreme weather conditions associated with depressions and anticyclones.

Answers on p. 123

How extreme weather conditions develop   Revised ☐

Tropical cyclones

Tropical cyclones are powerful storms that develop over warm oceans, between latitudes 7° and 20°.

Form and process

Tropical cyclones are areas of low pressure that are circular in plan. In the Northern Hemisphere surface winds circulate anticlockwise, spiralling in towards the low-pressure centre. At the centre of the storm is the **eye**. This warm core, around 50 km in diameter, is calm and often cloud-free. Within the eye air sinks towards the surface, preventing cloud formation. Bordering the eye is the **eye wall**, a ring of tall thunderstorm clouds that produce heavy rain and fierce winds.

Throughout, most of the storm air rises vertically, creating low pressure at the surface and high pressure aloft. At upper levels the air diverges, complementing the **convergence** near the surface.

Development of tropical cyclones

A number of favourable weather and ocean conditions are needed for the development of tropical cyclones:

- high humidity, supplying abundant water vapour
- light winds to allow vertical cloud development
- convergent winds at low levels and divergent winds aloft
- sea surface temperatures (SSTs) of at least 26–27°C and a deep warm-water layer of 60–70 m (to prevent cold water rising to the surface and killing the system).

Table 3.3 Researching the development of hurricanes

Source	Information
National Hurricane Center (www.nhc.noaa.gov)	Provides detailed information on the development and tracks of named hurricanes.
Rainfall radar data (www.weather.gov)	Regular updates of rainfall patterns from radar.
Satellite images (www.sat.dundee.ac.uk)	Further information on tracking and development of hurricanes available from satellite images.
Weather measurements (www.ndbc.noaa.gov)	Weather data recorded by automatic weather stations on buoys moored at sea.

Snow and ice

Blizzards

Blizzards develop when heavy snowfall is driven by strong winds. Such extreme weather is most often caused by deep depressions and frontal systems during spells of sub-zero temperatures.

Cold spells

The winter of 2009–2010 in the UK was remarkable for a protracted cold spell lasting from mid-December to mid-January. It was was the longest period of severe winter weather experienced in the UK since 1963, which was the coldest winter of the twentieth century. At Manchester the average daily temperature between 17 December and 9 January was −2.5°C.

Prolonged cold spells in western Europe are usually caused by **blocking anticyclones**. In 2009–2010 the normal westerly circulation over the UK broke down and high pressure from eastern Europe and Siberia extended westwards, flooding western Europe with freezing arctic air.

Drought

Droughts result from prolonged periods of abnormally low rainfall. They develop slowly and lead to water shortages that may last for weeks or even months.

> **Typical mistake**
> Drought is not just caused by low rainfall. Excessive moisture losses from evaporation and transpiration contribute to drought.

Causes of drought

The main meteorological cause of drought in western Europe is persistent high pressure, which blocks the passage of low-pressure systems that normally bring precipitation.

Droughts are common in the tropics and sub-tropics when seasonal rains fail. In south Asia the failure of the monsoon rains in 2009 led to drought and food shortages. Similar events have occurred frequently in Sub-Saharan Africa over the past 40 years when seasonal rains have failed.

Extreme impacts

Factors affecting the impact of extreme weather
Revised

Extreme weather becomes hazardous when it has adverse impacts on people and society. These impacts may include loss of life, injury, and damage to property and infrastructure. Three factors influence the severity of the impact of extreme weather:

- the scale and magnitude of the hazardous event
- the number of people exposed to risk in the affected area
- the vulnerability of a population and its level of preparedness

> **Enquiry question**: What are the impacts of extreme weather on people, the economy and the environment?

Scale and magnitude

Tropical cyclones

The more powerful and intense a weather event, generally the more severe its impact. The Saffir-Simpson scale grades hurricanes from 1 (weakest) to 5 (strongest). Hurricane Katrina, a powerful (category 5) hurricane devastated the Gulf coast of Louisiana in August 2005. It killed an estimated 1,350 people, caused damage in excess of US$100 billion, and was the costliest natural disaster in US history.

> **Examiner's tip**
> The human impact of extreme weather often depends on its geographical extent. Tornadoes are confined to localised areas, whereas hurricanes and floods are much more extensive.

> **Examiner's tip**
> Descriptions of extreme weather events must be supported with actual examples.

Tornadoes

Tornadoes are destructive but localised weather phenomena. Their impact depends on their intensity: winds in tornadoes can reach $500\,km\,h^{-1}$, making them the most violent storms on the planet. The Fujita tornadic damage scale (F1–F5) provides an indication of the power of tornadoes.

Exam practice answers and quick quizzes at **www.hodderplus.co.uk/myrevisionnotes**

Extreme rainfall

In the UK, long sequences of deep Atlantic depressions can result in exceptional rainfall events. This happened in late autumn 2009. November 2009 was the wettest month on record in the UK, with rainfall averaging 217 mm across the whole country. The heaviest rain, intensified by the Lakeland hills, fell in Cumbria between 17 and 20 November. The rivers Derwent and Cocker burst their banks and caused widespread flooding throughout the county.

Population at risk

The impact of extreme weather hazards partly depends on the number of people living in areas at risk. In 1970 and 1991, tropical cyclones caused huge loss of life in the Ganges-Brahmaputra delta of Bangladesh and India (around 600,000 deaths). The delta is home to between 125 and 140 million people. The majority of its inhabitants live within 1 m or so of sea level.

Vulnerability and preparedness

The risks posed by extreme weather also depend on the vulnerability and preparedness of a society. In general, populations in poor countries are most vulnerable to extreme weather. In the USA, vulnerability to hurricanes is reduced by:

- comprehensive monitoring and tracking of hurricanes by the US National Hurricane Center
- information relayed to weather forecasters and broadcast to Gulf-coast residents at risk through TV, radio and online
- planned evacuation by US state governments and federal agencies emergency relief plans
- hard engineering structures such as levées and flood gates protecting the areas of highest risk

> **Now test yourself**
>
> 4 What scales are used to measure the magnitude of (a) tropical cyclones, (b) tornadoes?
>
> **Answer on p. 124**
>
> Tested ☐

> **Examiner's tip**
>
> The importance of vulnerability and preparedness to the impact of extreme weather events can be assessed by comparing hazards of similar magnitude in regions of contrasting economic development.

Fieldwork and research

Revised ☐

Fieldwork and research should be conducted into an immediate and disastrous weather event (Table 3.4), a subsequent additional hazard (Table 3.5), and a longer-term trend or condition (Table 3.6).

Table 3.4 Investigating an immediate and disastrous weather event: the tornado disaster in Mid-West USA, 2–3 March 2012

Secondary sources
Scientific American articles: **www.scientificamerican.com**
New York Times articles: **www.nytimes.com**
Los Angeles Times articles: **www.latimes.com**
Newspaper reports provide immediate coverage of the destructive impact of tornadoes.
Primary sources
National Climate Data Center: **www.ncdc.noaa.gov**
Storm Prediction Center: **www.spc.noaa.gov**
Provide data on monthly tornado counts, the weather conditions that led to the 2–3 March outbreak and maps of damage tracks of tornadoes.
Earth Sky: **www.earthsky.org**
Provides photos of storm damage, radar animations of storm tracks, maps of storm tracks.

Table 3.5 Investigating localised flooding: flash floods in Hebden Bridge, Calderdale, 23 June and 9 July 2012

Secondary sources
Descriptions of flood impacts and photos at: **www.dailymail.co.uk**
www.guardian.co.uk
www.todmordennews.co.uk
Videos of floods at: **www.hebdenbridgetimes.co.uk**
www.itv.com/news
www.bbc.co.uk/news/uk
www.sky.com
Flood maps and flood warnings: **www.environment-agency.gov.uk/homeandleisure/37837.aspx**

Primary sources

Centre for Ecology and Hydrology for catchment and flow data, catchment descriptions and factors affecting runoff in Calderdale: **www.ceh.ac.uk**

Unofficial historic rainfall statistics (daily, monthly rainfall and rainfall charts for 24 hours) for the flood period at: **www.wunderground.com/weather-forecast/UK.html**

River level data on Hebden Water and River Calder at: **www.environment-agency.gov.uk**

Fieldwork might include questionnaire surveys on (a) the extent of flooding and (b) the impact on residents, businesses, infrastructure and transport.

Examiner's tip

Before investigating a localised weather event you must be sure that sufficient primary and secondary data sources are available. Remember that official data may not be available until several months after the event.

Examiner's tip

Look at the websites for primary and secondary data in Tables 3.4 and 3.5 with a view to using similar data sources for your own investigations.

Table 3.6 Investigating the impact of drought in England, 2011–2012

Secondary sources

The environmental impact of drought at: **www.naturalengland.org.uk**

The economic and social impact of drought (e.g. on farming) at national newspaper websites: **www.guardian.co.uk**
www.telegraph.co.uk

The impact of drought on farming and the environment at: **www.defra.gov.uk**

Videos of impact of drought on farming: **www.bbc.co.uk/news**

Primary sources

Explanations of drought, including rainfall maps for UK November 2010–February 2012 (as per cent long-term average) and comparisons with 1975–1976 droughts, at: **www.metofficenews.wordpress.com**

Maps of reservoir and groundwater levels and river flow (compared with long-term average) at: **www.environment-agency.gov.uk**

Rainfall in Anglian Water region July 2011–June 2012 at: **www.anglianwater.co.uk**

Fieldwork interviews with groups affected by drought. Assessments of drought by measuring moisture content of soils, monitoring water levels in reservoirs, recording daily rainfall totals and maximum temperatures, etc.

Increasing risks

Evidence of more frequent extreme weather events

Revised

Since the mid-1970s there has been a global increase in the frequency of extreme weather events (see Unit 1, 'Global hazard trends', p. 7). This trend, together with global population growth and poor land management, has increased the risks to human populations.

Enquiry question: How are people and places increasingly at risk from and vulnerable to extreme weather?

Natural causes

More extreme and more frequent high temperatures result in an increasing number of heat waves and related hazards such as droughts and wildfires. Examples include the record-breaking 2003 heat wave in Europe, and the 2012 heat wave in the USA.

Extreme rainfall and flooding also occur with more frequency. Because global temperatures have increased, the warmer atmosphere contains more moisture. One outcome is that torrential rainfall events are more frequent. In 2012 the UK recorded its wettest-ever April–June, while June 2012 was the wettest June since records began in 1910.

Typical mistake

Although extreme weather hazards are occurring more frequently and with greater intensity, this fact alone does not explain their increased impact. The importance of human factors, such as population growth and land-use mismanagement (often neglected), must also be considered.

Human causes

The growth of population and settlement on floodplains and deltas increases exposure to and the risks of flooding. The huge impact of Hurricane Katrina in 2005 was not just due to the magnitude of the storm (category 5). Also significant were:

Exam practice answers and quick quizzes at **www.hodderplus.co.uk/myrevisionnotes**

- the fact that 11–12 million people inhabit the coastal counties between Louisiana and Florida along the Gulf of Mexico
- rapid population growth, which had taken place in these communities in the previous 50 years or so, i.e. a three-and-a-half times increase between 1950 and 2004, and a 7% rise between 2000 and 2004.

Land management

- The Hurricane Katrina disaster was exacerbated by subsidence in the Mississippi delta caused by natural gas extraction, and levées that 'walled in' the Mississippi River, preventing annual floods and natural deposition.
- Poor land management, such as deforestation, increases runoff and the flood risk. The floods caused by Hurricane Mitch in Honduras and Nicaragua in 1998 were triggered by (a) torrential rain, and (b) extensive deforestation of upland catchments by farmers, loggers and mining operations.
- Urban growth covers the ground with impermeable surfaces that accelerate runoff and raise peak river flows. The encroachment of towns and cities into the countryside contributes to the loss of wetlands and lakes — natural stores of water that help to prevent flooding.

> **Typical mistake**
>
> It is impossible to say that specific extreme weather events, such as the 2005 hurricane season in the USA, are due to global warming and climate change. What scientists can say with confidence is that the general trend towards weather extremes is related to global warming and increased GHG emissions.

> **Typical mistake**
>
> Remember that global warming cannot explain a single extreme weather event, such as a flood or a drought. However, global warming can be used to explain general patterns and trends in extreme weather.

Fieldwork and research

Revised

Research into a flood event on a small stream or part of a river catchment will rely on fieldwork observation and a range of primary and secondary data derived from maps and internet sources (Table 3.7).

Table 3.7 Investigating a flood event in a small river catchment

Background	Catchment area, watersheds, altitude, slopes, geology, vegetation from 1:25,000 and 1:50,000 OS maps and 1:50,000 Geological Survey maps.
Land use	Land use descriptions and the factors affecting runoff for gauged catchments are available at the Centre for Ecology and Hydrology website: **www.ceh.ac.uk**
Rainfall	Synoptic charts showing the weather situation at the time of the flood, available at: **www.metoffice.gov.uk/public/weather/surface-pressure/**
	Rainfall radar images are also available on the Met Office website at: **www.metoffice.gov.uk/public/weather/observations/?tab=map&map=Rainfall**
	Hourly rainfall amounts are available for local weather stations though they are not archived: **www.metoffice.gov.uk/weather/uk/observations/**
	Unofficial historical rainfall data can be obtained for local areas from automatic independent weather stations: **www.weatherstations.co.uk/aws_map.htm**
	Actual rainfall amounts during the flood event can be compared with long-term averages, available for weather stations in the UK at: **www.metoffice.gov.uk/climate/uk/stationdata/**
River flows	The National Water Archive provides data on average daily river flows for gauging stations throughout the UK: **www.ceh.ac.uk/data/nrfa/data/search.html**
Areas vulnerable to flooding	The Environment Agency flood maps for England and Wales at a 1:40,000 scale. The maps show flood defences and the areas at risk at 1% and 0.1% probability. **www.environment-agency.gov.uk/homeandleisure/37837.aspx**
Fieldwork	Historic flood heights from markers on bridges and buildings.
	Recent flood heights from rubbish lines on floodplains and debris caught by bankside vegetation.
	Photographs and interviews with residents to determine the extent and depth of floods.
Land use change	Information on land use changes in a catchment by comparing field surveys with historic maps, or comparing modern OS maps with maps of similar scale published in the twentieth century.

> **Examiner's tip**
>
> Look at the websites listed in Table 3.7. Examine how the data available on each website could be used to investigate local flooding, their value to this type of study, and their limitations.

Managing extreme weather

Hurricanes hazards in the USA

Short-term responses

- Weather forecasts, updates and warnings are issued by agencies such as the US National Weather Service and the National Hurricane Center.
- Early warning of imminent storms may trigger mass evacuations, such as New Orleans in 2005 (Hurricane Katrina) and in Houston in 2008 (Hurricane Ike).
- People in hurricane hazard zones are encouraged to help themselves. The National Oceanic and Atmospheric Administration (NOAA) has a hurricane preparedness website.

Long-term responses

- Strengthening levées, sea walls, storm shelters and breakwaters to protect people and property.
- Reducing the risk of flooding from storm surges by land-use zoning, conserving wetlands and not interfering with natural sedimentation processes.
- The Federal Emergency Management Agency (FEMA), whose mission includes responding to, aiding recovery from, and helping to mitigate natural disasters.

> **Enquiry question**: How can we best respond to and cope with the impacts of extreme weather?

> **Examiner's tip**
>
> Short-term responses to hurricane (and other) hazards mainly comprise warnings, evacuation and emergency planning to deal with the immediate aftermath. Long-term responses concern planning to mitigate the impact of future hazards and disasters.

Table 3.8 Investigating the short- and long-term responses to hurricanes in the USA

Short-term responses	
Geostationary satellites	Free satellite images of the Gulf coast and Caribbean, updated at 12-hour intervals, are available at: **www.sat.dundee.ac.uk**
Storm tracks	Forecasts are issued by the National Hurricane Center (NHC): **www.nhc.noaa.gov/** Historical hurricane tracks can be found at: **csc.noaa.gov/hurricanes/#**
Ships and buoys	Provide air temperature, sea surface temperature, wind speed, wind direction, pressure and humidity data. See: **www.ndbc.noaa.gov/maps/Florida.shtml**
Hurricane warnings	Current hurricane watches and hurricane warnings are available at the NHC website (see above).
Evacuation	States exposed to hurricane hazards (e.g. Texas, Louisiana) have their own evacuation strategies and plans, e.g.: **www.txdps.state.tx.us/dem/downloadableforms.htm**
Longer-term responses	
Hard engineering	Protection against storm surges by strengthening/building levées, sea walls and breakwaters. Storm shelters and buildings on stilts.
Soft engineering	Removing obstacles to natural sedimentation in deltas and low-lying coasts (e.g. levées). Promoting the health of wetlands (e.g. salt marshes) that mitigate the impact of storm surges.
Personal preparedness	Educating populations at risk on actions to take to protect themselves and their property from hurricane hazards: **www.nhc.noaa.gov/prepare/ready.php**
Land use management	Restricting development in coastal areas susceptible to storm surges. Abandoning urban areas at or below sea level where risks are unacceptably high (e.g. some suburbs in New Orleans).
Emergency aid	FEMA assists regions affected by natural disasters. The role of FEMA can be investigated at: **www.fema.gov/**

River floods in the UK

Five million people in England and Wales are at risk from flooding by rivers and the sea.

Short-term responses

- The Environment Agency operates a flood warning system. Warnings can be accessed by telephone, mobile phones, e-mail and texts.
- People in areas at high flood risk can evacuate to higher ground, move belongings to upper storeys, install temporary flood gates and sandbags, etc.

Long-term responses

- Flood insurance to compensate householders for damaged property and repairs.
- Flood hazard maps and assessment of risks from floods.
- Land use planning — restricting development in areas of high flood risk, such as floodplains.
- Soft engineering to reduce rainfall:runoff ratios, e.g. afforestation, conservation of wetlands, allowing rivers to flood naturally onto their floodplains.
- Hard engineering structures to regulate river flow and/or confine floodwaters to river channels.

> **Typical mistake**
>
> Flooding not only occurs when rivers burst their banks or storm surges overwhelm sea defences, it also occurs when drains and sewerage systems cannot cope with extreme rainfall and when, after heavy and prolonged rain, the water table rises and groundwater reaches the surface.

Table 3.9 Investigating the short- and long-term responses to floods in the UK

Short-term responses	
Flood warnings	The Environment Agency's flood warning system and the action people should take, and warnings currently in force, are available at: **www.environment-agency.gov.uk/homeandleisure/floods/34678.aspx**
Personal preparation	Advice on preparations for a flood event can be found at: **www.environment-agency.gov.uk/homeandleisure/floods/31624.aspx**
Longer-term responses	
Hard engineering	Investigate current hard engineering flood alleviation schemes, e.g. Banbury and Cockermouth: **www.environment-agency.gov.uk**
Soft engineering	Washland and other schemes that allow temporary storage of floodwaters: **archive.defra.gov.uk/environment/flooding/documents/manage/jointstment.pdf**
Land use management	Strategies to alleviate flooding in the Derwent catchment, Yorkshire, which include non-structural approaches and land use management: **www.ceh.ac.uk**
	http://knowledge-controversies.ouce.ox.ac.uk/ryedaleexhibition/

New technologies
Revised

New technologies can improve community preparedness for extreme events such as tropical cyclones and floods, and thus reduce their impact.

Weather forecasting

The use of satellites and automated weather stations has greatly increased the accuracy of weather forecasts over the past 30–40 years. Extreme weather phenomena (e.g. tropical cyclones) are monitored by satellite images, doppler radar, radio-sondes and aircraft. This information is used by agencies to issue forecasts and warnings of extreme weather events.

Flood forecasting

In the UK the Environment Agency issues flood warnings. They rely on data from:

- weather forecasts
- river catchment models
- historical hydrological data that relate precipitation events to river discharge

Drought-resistant crops

Scientists have improved crop yields in drought-prone areas by developing new crops that are either more resistant to drought or that mature earlier. This has been achieved by (a) selective plant breeding and (b) genetic engineering. Important advances have been made in staple dryland crops grown in LEDCs such as millet and sorghum.

Managing drought

Water management and adapting farming techniques are options for tackling the problem of drought.

> **Examiner's tip**
>
> The geographical impact of improvements in technology in mitigating weather-related disasters has been uneven. Economic development appears to be a major influence in the effectiveness of hazard mitigation.

Water management in California

Most of California has a semi-arid climate; drought is commonplace and water demand from agriculture and domestic consumers is high. One-third of California's water comes from the Sierra Nevada mountains and derives from the snowpack that forms each winter.

California's vulnerability to drought is reduced by the state's elaborate water infrastructure. This comprises surface reservoirs, boreholes and aqueducts. Aqueducts transfer water from the Owens Valley in eastern California to Los Angeles, and from Lake Havasu on the Colorado River to southern California.

Because California is running out of water and its population is due to increase by 10 million in the next 15 years, water conservation has become an urgent issue. Lower outdoor consumption, including replacing lawns with desert plants and improving the efficiency of irrigation, are essential if water usage is to be sustainable in future.

Adapting farming techniques in southeast England

Climate change is likely to mean warmer and drier summers in southern Britain. Increasing aridity poses a challenge to farming. Possible responses include:

- extending the area of irrigated cropland
- **trickle irrigation**, which is more water efficient and produces higher yields than spray irrigation
- **zero tillage**, which is planting crops in unploughed soils. Zero tillage reduces moisture losses to evaporation
- tree shelterbelts reduce moisture losses to **evapotranspiration**
- new crops such as maize and soya that are better adapted to drier and warmer climates than temperate crops such as wheat and barley

> **Examiner's tip**
>
> Think of water management as matching supply with demand, ensuring the sustainability of supply, and minimising the impact of water sourcing and transfer on the environment.

> **Now test yourself**
>
> 5 State four ways technological innovations could help farmers cope with drought in the UK.
>
> **Answer on p. 124**
>
> Tested ☐

> **Exam practice**
>
> **1** Study Figure 3.1 on p. 84.
>
> **(a)** Describe the potential extreme weather hazards associated with the features on the synoptic chart. [10]
>
> **(b)** Describe the fieldwork and research you undertook to investigate an extreme weather hazard. [15]
>
> **(c)** Using examples, explain how technology can reduce the impacts of extreme weather hazards. [10]
>
> **Answers and quick quiz 3 online**
>
> Online ☐

4 Crowded coasts

Competition for coasts

Coastal physical environments
Revised

Geology, geomorphology and ecology combine to create four main types of coastal physical environment:

- hard rock upland coasts, dominated by erosional features
- beach environments, of sand and shingle
- mudflats and salt marshes
- coastal dunes

Hard rock upland coasts

Upland coastlines made of resistant rocks support a range of erosional landforms. They are the result of wave erosion and **sub-aerial processes** and the retreat of cliffs inland. As cliffs retreat they create a distinctive sequence of residual landforms: **caves**, **arches**, **stacks** and **shore platforms**.

Structure and **lithology** influence rates of cliff recession. Wave action abrades rocks in the intertidal zone, and quarrying by waves causes cliff collapse along major joints. Sub-aerial processes such as weathering and mass movement also contribute to cliff recession.

Beach environments

Beaches are accumulations of sand and shingle in the shore zone (Figure 4.1).

Beach profiles are the cross-sectional shape of beaches between the mean high- and low-water marks. Two factors determine the shape of beach profiles: sediment size and wave type.

In planform, beaches are either **swash-aligned** or **drift-aligned**. Swash-aligned beaches are more or less straight (or crescent-shaped) and develop where waves are fully **refracted**. Drift-aligned beaches, which include **spits** and **barrier beach islands**, develop on open coastlines where waves are rarely fully refracted. As a result sand and shingle are transported along the coast by **longshore drift**.

Mudflats and salt marshes

Mudflats and salt marshes are features of low-energy coastlines where wave action is weak and tidal currents deposit fine sediment. Sediment accretion leads to the growth of mudflats and salt marshes.

Vegetation has a key role in the development of salt marshes.

- Pioneer species such as cord grass and glasswort colonise the mudflats. These plants slow the movement of water and encourage rapid sedimentation (1–2 cm per year). Their roots also help to stabilise the mud.
- Through accretion, mudflats grow in height causing the period of inundation on each tidal cycle to decrease. At the same time salinity declines and a continuous cover of vegetation is established. The salt marsh is dominated by specialised plants such as sea rush, sea aster and sea lavender.

> **Enquiry question**: Why is the coastal zone so favoured for development?

> **Structure** refers to the physical characteristics of rocks, such as joints, faults and bedding.
>
> **Lithology** is the chemical composition of rocks.

> **Examiner's tip**
>
> Think of landforms such as caves, arches, stacks and shore platforms as temporary features that indicate the stage of retreat of cliffs through erosion, weathering and mass movement.

> **Now test yourself**
>
> 1 Name four erosional landforms of hardrock upland coasts.
>
> **Answer on p. 124**
>
> Tested

> **Typical mistake**
>
> Beaches are classified according to their shape or planform rather than the processes that built them (i.e. swash- or drift-aligned). This means that beaches such as spits, tombolos and offshore bars could be either swash- or drift-aligned.

Coastal dunes

Coastal dunes, a feature of lowland coastlines, develop where (a) sand supply is abundant, (b) offshore gradients are shallow, (c) prevailing winds are onshore and (d) there are extensive backshore areas where sand can accumulate (Figure 4.1). Classic dunes consist of a series of three or four linear ridges that run parallel to the coastline, separated by depressions known as **slacks**.

Mature dunes form when vegetation provides an obstacle to blown sand. Dune plants, such as **marram grass**, thrive when submerged by sand. Burial stimulates growth, which in turn encourages further deposition. Deposition also occurs because plants reduce wind speeds near the ground surface. Thus by anchoring the sand, vegetation plays a key role in dune formation.

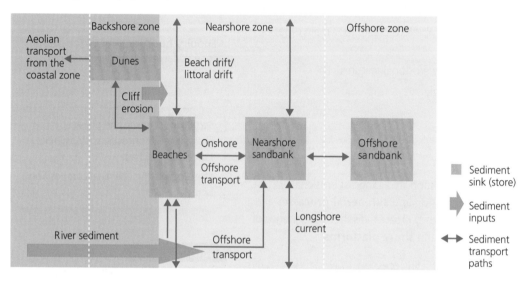

Figure 4.1 Movement and stores of sediment in the coastal zone

Now test yourself

Tested

2 What physical factors influence the formation of sand dunes?

Answer on p. 124

Rapid population growth in coastal environments

Revised

Coastal environments are highly attractive to human populations. In the coterminous USA, 53% of the population lives in coastal counties that occupy just 17% of the total land area. Several factors account for the popularity of coastal regions:

- deep water ports and harbours (most mega cities are ports), access to offshore oil and gas reserves, fish stocks, etc. that attract economic activities

- fertile soils derived from alluvial and tidal sediments, and fresh water for irrigation (e.g. in deltas). A combination of favourable climate and hydrological conditions, and fertile soils allow multiple cropping

- beaches, the sea and sunnier coastal climates attract outdoor recreation and tourism. In MEDCs the coast is also a magnet for retirees.

Now test yourself

3 What reasons explain the rapid growth of population in many coastal areas?

Answer on p. 124

Tested

Fieldwork and research

Revised

Fieldwork might investigate migration patterns, occupations, economic activity, origins of tourists and types of tourism in coastal areas. Observational surveys of guest houses, hotels, caravan and camping sites, retailing, fishing, etc. could provide information that allows connections to be made with environmental resources.

Table 4.1 The growth of contrasting crowded coasts

Investigator	Cornwall	Costa del Sol
Traditional economic activities and economic regeneration	Fishing, mining, etc.	www.andalucia.com/spain/statistics/home.htm
Economic importance and growth of tourism	www.ons.gov.uk/ons/dcp171776_239218.pdf	www.andalucia.com/spain/statistics/population.htm Statistics 1991–2010 on population change, house building, economic activity, airport passenger numbers, and tourism at regional and municipality scales.
Demographic change	www.neighbourhood.statistics.gov.uk www.ons.gov.uk/ons/publications/re-reference-tables.html?edition=tcm%3A77-231847 Information on population change, age structure, migration and household composition 2001–2010.	International retirement migration: landscaperesearch.livingreviews.org/open?pubNo=lrlr-2010-2&page=articlese4.html Growth of tourism on Costa del Sol: www.visitcostadelsol.com/discover-costa-del-sol/general-info/history-of-the-costa-del-sol
Changing land use and growth of the built environment	Comparison of recent OS maps with historic OS maps for the period 1930s, 1940s and 1950s.	Coastal urban sprawl in Spain and elsewhere in Europe: www.scribd.com/doc/32841769/Urban-Sprawl-in-Europe-The-Ignored-Challenge

Coping with the pressure

Pressure for space and land use zoning

Revised

Tourism development

In the past 50 years rapid urbanisation, driven by package holidays, cheap air travel and **mass tourism**, has occurred along the Mediterranean coast of southern Spain.

Largely uncontrolled, the environmental costs of this development have been considerable. High-rise hotels have encroached to within 20 m of the shoreline. In the past planning was non-existent. Resort growth outpaced the investment in essential infrastructure and beach pollution became widespread. As tourism grew the provision of sustainable water resources in the dry Mediterranean summer proved challenging.

In the 1990s and early twenty-first century demand for housing by second-home owners and retirees from northern Europe brought a second building boom. *Urbanizaciónes*, often built illegally, sprawled across sites set back from the coast where land was available at relatively low cost.

Zonal structure

Coastal resorts in Spain typically have a zonal structure. Premium locations, with access to the beach and sea views, form a strip along the sea front. High-rise hotels dominate this zone, where competition forces up rents and land values. The resort strip also supports concentrations of tourism-related activities such as bars, restaurants, clubs and shops. Behind the resort strip are semi-circular zones of low-rise buildings. These zones accommodate local residents and workers in the tourism industry. Housing is often traditional and low quality, streets are narrow and congested with parked cars.

Beyond the built area is an extensive rural zone where traditional farming may still survive. Major roads cross this zone connecting neighbouring resorts, and villages may offer specific tourist attractions such as historic churches, castles, local crafts and trails.

Enquiry question: How do various coastal developments create competition and conflict? How can these pressures be resolved?

Examiner's tip

Learn how rapid and unplanned growth of mass tourism in countries such as Spain and Greece has often conflicted with environmental protection, degraded natural resources and swamped local, traditional cultures and lifestyles.

Economic pressures on coastlines produce conflicts between development and conservation. Developers emphasise the benefits of economic growth and new employment. Conservationists point to the overuse of resources, pollution and the destruction of wildlife habitats.

Table 4.2 Researching the pressures and conflicts between development and conservation in coastal environments

Destruction of mangroves in Bangladesh	Coastal mangroves have great ecological importance: they are biodiverse and highly productive, and vital nurseries for young fish, crustaceans and molluscs. The mangroves are threatened by overuse and deforestation (they provide charcoal, timber and fuelwood). **www.preventionweb.net/files/8203_SalinityandMangrove.pdf** **www.rtcc.org/nature/satellite-photo-destruction-of-the-mangrove-forest-of-the-sundarbans/**
Shrimp farming in coastal Bangladesh	Shrimp farming is a leading activity on the southeast coast of Bangladesh, worth more than US$300 million per year. Its success has caused significant damage to the environment. Shrimp farms have replaced mangrove forests. Polluted water, enriched with fertilisers from shrimp ponds, is released to the environment. **www.tropentag.de/2008/abstracts/full/461.pdf**
Foreran links golf course and impact on coastal dunes	A new links golf complex (golf course, hotel, timeshare flats) was completed in 2012 on the Foveran dunes in Aberdeenshire. The development aroused strong opposition from conservationists because the dunes are (a) a conservation area (SSSI), (b) of great ecological and geomorphological interest and (c) extremely fragile. **www.scottishwildlifetrust.org.uk/what-we-do/policy-and-campaigns/**

Coastal development: economic benefits and environmental costs

Analysis of the economic impacts and the environmental costs of a development often use 'tools' such as **cost–benefit analysis** (CBA) and **environmental impact assessment** (EIA).

Severn barrage tidal power scheme

The projected Severn barrage is a scheme designed to harness the tidal power of the Severn estuary (Figure 4.2). The huge tidal range in the estuary would be exploited to drive turbines and generate large amounts of 'green' energy. The project could provide considerable economic benefits but only at the expense of major environmental costs.

> **Examiner's tip**
>
> Fieldwork and research should focus on an environmental issue that polarises opinion. Remember that arguments are often finely balanced, and that decisions are based on judgements that aim to maximise the benefit to society.

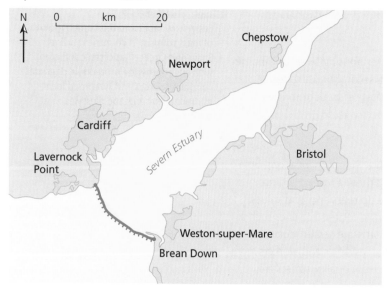

Figure 4.2 The proposed Severn barrage

Environmental impact assessment

Table 4.3 sets out the positive and negative environmental impacts of the Severn barrage.

Table 4.3 Environmental assessment impact of the Severn barrage

Negative impacts
• Loss of saltmarsh and mudflat habitats — important feeding grounds for birds (waders and wildfowl). The estuary is a conservation wetland (Special Protected Area, Special Area of Conservation, Ramsar, SSSIs).
• Migrating fish, such as salmon, shad and twaite, would be severely affected, with loss of spawning habitats.
• Sediment flows in the estuary would be disrupted, with uncertain consequences.
• The Severn tidal bore — one of the largest in the world — would either be reduced in size or lost.
• Problems with the dilution and dispersal of sewage waste in the absence of strong tidal currents.

Positive impacts
• Generation of renewable energy and reduction of carbon emissions in the UK (1 million tonnes per year) helping the UK meet its international climate change obligations.
• Some bird species (e.g. dunlins) could benefit from more stable water levels in the estuary.
• Studies of the impact of the world's only tidal power station, La Rance in Brittany, have suggested that tidal power schemes can increase biodiversity, and improve water quality and oxygen levels.

Cost–benefit analysis (CBA)

CBA provides a method to compare the costs and benefits of the Severn barrage. Benefits are broadly defined as improvements in human well-being; costs are decreases. If benefits exceed costs, the project is beneficial as a whole, and is likely to be approved. CBA is most effective where costs and benefits can be quantified in monetary terms.

Table 4.4 Cost–benefit analysis of the impact of the Severn barrage

Benefits
• Will generate 5% of the UK's electricity needs — equal to three nuclear power stations.
• A renewable form of energy, free from carbon emissions. More predictable than wind and solar power.
• Will help to fill the UK's 'energy gap' as many older nuclear power stations are decommissioned over the next 5–10 years.
• Will have a life span of up to 150 years.
• At construction peak could create 35,000 jobs and 10,000 permanent jobs locally.
• Will provide opportunities for new leisure activities, including water sports, bird watching and angling.
• Will strengthen flood defences around the Severn estuary; the barrage will also act as a barrier to storm surges.
• Could provide new road and rail links across the estuary.

Costs
• Huge cost of the project — around £20 billion.
• Disruption to shipping and ports such as Avonmouth, Cardiff, Newport and Sharpness.
• Beaches at Weston-super-Mare could be degraded, thereby damaging tourism.
• Negative impact on local fishing and the extraction of aggregates from the estuary.
• Extra flood defences needed to protect against excessive runoff in the Severn, Wye and Usk catchments.
• Negative impacts on the environment: loss of habitats (saltmarshes, mudflats), SSSIs, reduction in bird and fish populations, loss of Severn tidal bore, problems of sewage dispersal (and other effluents).
• Unpredictable impacts on sediment movement in the estuary.

> **Typical mistake**
>
> CBA is far from straightforward. It runs into problems when development schemes have an environmental dimension. For example, it is impossible to place a monetary value on rare and endangered species such as great crested newts or fly orchids threatened by development.

Conflict matrices

Conflict matrices assist analysis of opposing views and objectives of interested parties and decision-makers in complex economic, social and environmental issues such as the construction of the Severn barrage (Figure 4.3). In Figure 4.3, quadrants with a cross indicate a potential conflict between the interested parties.

	1	2	3	4	5	6	7	8	9
1		x			x	x			x
2			x	x			x	x	
3					x	x			x
4					x	x			x
5							x	x	
6							x	x	
7									x
8									x

1 = conservationists, 2 = water sports enthusiasts, 3 = fishermen, 4 = aggregate extractors,
5 = residents at risk from flooding, 6 = local people seeking employment and economic development agencies,
7 = port operators, 8 = beach tourists, 9 = energy companies.

Figure 4.3 Conflict matrix of interested parties in the Severn barrage scheme

Increasing risks

Coastal development and erosion and flooding

Revised

The growing level of coastal development and climate change are increasing the risks of coastal erosion and flood hazards.

Enquiry question: How is coastal development increasingly at risk from, and vulnerable to, physical processes?

Accelerated coastal erosion

Rates of coastal erosion will accelerate in future as global warming:

- increases the strength of the jet stream, raises sea surface temperatures and generates storms of greater frequency and intensity
- raises sea level, eroding dunes, salt marshes and mudflats, and extending the shore zone across which erosional processes operate

Rising sea level

Sea level rose by nearly 20 cm between 1900 and 2000. This was due to the rapid melting of glaciers and ice sheets, and the **thermal expansion** of the oceans caused by global warming. Current forecasts suggest an average rise in sea level of 40 cm by the end of the twenty-first century. This could spell disaster for countries such as Bangladesh, where 37% of the land area is less than 3 m above sea level.

Rapid erosion along vulnerable coasts

Coastal erosion in eastern England

Eastern England has some of the most rapidly eroding coastlines in the world. At Holderness in east Yorkshire, cliff retreat averages 2 m per year. Since Roman times the coastline has receded by more than 5 km (Figure 4.4). Similar rates of erosion occur along the East Anglian coast, most notably at Happisburgh and Overstrand in Norfolk. Rapid erosion is due to:

- geology and relief: low cliffs made of soft materials such as clay, sand and gravel
- an absence of broad beaches that might otherwise dissipate wave energy
- longshore drift, which transports eroded material (that might otherwise build beaches) along the coast

Coastal erosion mainly results in the loss of farmland. However, at Ringborough and North Cowden in Holderness and at Happisburgh, erosion has forced the abandonment of farms and residential properties in recent years.

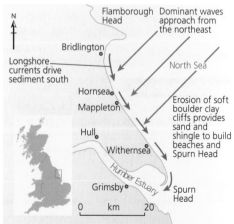

Figure 4.4 Coastal erosion at Holderness

Exam practice answers and quick quizzes at **www.hodderplus.co.uk/myrevisionnotes**

The impact of rising sea level

Rising sea level increases the risks of flooding along lowland coastlines. London is just one of many world cities at risk. The Thames barrier, which protects the capital against exceptionally high tides and **storm surges**, is evidence of this risk.

Coastal flood risks are even higher along densely populated lowland coasts exposed to tropical cyclones and **tsunamis**. With higher sea levels, powerful tropical cyclones could become even more hazardous in future.

Tsunamis are a major threat in seismically active coastal regions. As sea level rises, the threat of tsunami hazards will increase. Nuclear power stations, sited on the coast, are especially vulnerable. Japan has 50 nuclear reactors, most of them located on the coast. The 2011 tsunami badly damaged the Fukushima nuclear plant, releasing large amounts of radiation and contaminating a wide area. The outcome was forced evacuation and the imposition of a 30 km exclusion zone around the plant.

> **Typical mistake**
>
> It is wrong to attribute current rapid coastal erosion to climate change. The Holderness coastline has been eroding rapidly throughout historic times. The important point to remember is that climate change will accelerate erosion in future, and this process is probably already underway.

> **Now test yourself**
>
> 4   What are storm surges and tsunamis and why might the risks from these hazards increase in future?
>
> **Answer on p. 124**
>
> Tested ☐

Fieldwork and research

Revised ☐

Fieldwork and research could investigate rates of coastal erosion and the resulting impacts on local communities.

● The present-day coastline could be surveyed through fieldwork and drawn on a large-scale map (1:10,000 or 1:2,500). Alternatively, a recent image of the coast from Google Earth could be used or an up-to-date 1:25,000 OS map.

● Mid- and late-nineteenth-century 6 inch OS maps provide historic evidence of the changing coastline. Aerial photos from the 1930s onwards provide further information.

● For east Yorkshire, the University of Hull has lists of erosion posts with their grid references and rates of erosion between 1951 and 2004. These data provide scope for investigating spatial differences in erosion rates along the Holderness coast.

   www.hull.ac.uk/coastalobs/hornsea/erosionandflooding/index.html

● Local newspapers have photos and reports of major erosional events in the past 30 years or so, e.g. large-scale cliff collapse, demolition of houses and farms threatened by erosion.

● Fieldwork observation will provide visual evidence of erosion such as damaged sea defences, pill boxes on beaches, etc.

● The impact of erosion on communities can be assessed by interviewing local residents.

> **Examiner's tip**
>
> Investigations should focus on a small stretch of coastline where rapid erosion is taking place *and* represents a hazard to local people (e.g. east Yorkshire, northeast Norfolk, Christchurch Bay).

Coastal management

Hard engineering

Revised ☐

The traditional response to problems of coastal erosion and flooding has been hard sea defences such as sea walls and groynes (Table 4.5). But increasingly, **hard engineering** is questioned on grounds of (a) its sustainability, and (b) its adverse environmental impacts.

> **Enquiry question**: How is coastal development adapting to new ideas and situations?

Table 4.5 Hard engineering sea defences

Sea walls	Sea walls are expensive to build and maintain and are only justified where important settlements and/or infrastructure are at risk. Sea walls stop erosion and prevent flooding but require costly maintenance. Because sea walls stop erosion they prevent beaches, mudflats and salt marshes from migrating inland and accelerate their erosion. In addition they reduce sediment input to the coastal system.
Revetments	Revetments are wood or rock barriers running parallel to the shoreline. They absorb wave energy and are cheaper than sea walls, but unsightly. They also limit access to beaches for visitors and tourists.

Groynes	Groynes are wood or rock barriers constructed perpendicular to the shoreline. By interrupting longshore drift and keeping beaches intact, they reduce wave erosion. However, they often starve beaches of sediment downdrift and accelerate erosion there.
Rip-rap	Rip-rap are boulders or concrete blocks placed at the foot of cliffs or in the backshore area of a beach. Although unsightly, they are a cheap and effective way of stopping erosion.
Gabions	Gabions are wire cages filled with rocks. When stacked, they form an effective and cheap defence against coastal erosion. They are, however, unsightly.
Barrages	Storm surge barriers such as the Thames barrier protect low-lying estuarine areas from flooding. The capital costs of building and operating them are huge and only justified where large populations and major investments are at risk.

Soft engineering approaches to coastal protection — Revised

On some coastlines, hard engineering has been replaced by **soft engineering**, which aims to work with, rather than resist, natural processes (Table 4.6). Soft engineering is cheaper than hard engineering, more environmentally friendly and more sustainable.

Table 4.6 Soft engineering approaches

Beach replenishment	Sand and shingle are imported to beaches to replace sediments lost to erosion. Problems occur where sand is mined offshore, disrupting the coastal sediment system.
Managed realignment	Existing hard defences may be dismantled or abandoned. The shoreline is set back, allowing the sea to flood areas previously protected. Salt marshes, mudflats and beaches develop and provide natural protection against erosion and flooding.
Dune regeneration	Sand fencing and marram planting help to stabilise dunes and reduce wind erosion. Boardwalks laid across dunes help to prevent trampling by visitors. Brushwood barriers around the high-tide level reduce wave erosion and encourage sand accumulation.
Marsh creation	The creation of salt marshes is a sustainable way of defending lowland coasts from flooding and erosion. Salt marshes raise land levels and dissipate the energy and erosive power of waves.

Examiner's tip

You need to learn the economic and environmental advantages and disadvantages of hard and soft engineering strategies on coasts. Of particular importance is the economic and environmental sustainability of each approach.

Fieldwork and research — Revised

The success of coastal defence schemes

Hard coastal defences could be assessed against the following criteria:

- the extent to which they stop erosion and/or flooding along a length of coastline
- their impact on adjacent coastlines unprotected by hard defences
- the extent to which defences disrupt the natural movement of sand and shingle (onshore/offshore/longshore)

Fieldwork

Possible fieldwork tasks include:

- observation and mapping of existing sea defences, their condition and maintenance
- evidence of erosion (e.g. seawalls being outflanked), assessment of current inputs of sediment through erosion and likely impacts on the coastal sediment budget
- evidence of longshore drift
- interviews with local residents to investigate the effectiveness of, and their attitudes towards, sea defences
- observation and mapping of erosion on adjacent coasts unprotected by hard defences

Exam practice answers and quick quizzes at **www.hodderplus.co.uk/myrevisionnotes**

- observation and recording of beach widths and beach heights in areas protected by groynes and sea walls

Secondary sources

- Old large-scale OS maps (1:10,000, 1:2,500, 6 inch) showing the coastline and former sea defences.
- Historic aerial photographs, postcards, etc. showing sea defences.
- Local newspaper reports on sea defences, and erosion and flooding hazards.
- Shoreline Management Plans (available online) for the coastline under investigation. Consideration should be given to strategies for sub-cells and management areas in the plan and their rationale, and the local sediment budget.
- Academic studies of the coast, e.g. east Yorkshire and University of Hull, Christchurch Bay and University of Southampton. An example of the latter is: **www.southampton.ac.uk/~imw/barteros.htm**

Strategies used to manage a high value coastal environment

Coastal dunes cover around 545 km² in the UK. They support rare and fragile ecosystems, easily degraded by human activities. Management of dune environments is essential for their protection and conservation. Various bodies are involved in conservation work, including local authorities, Natural England, wildlife trusts, DEFRA and the RSPB.

Investigation of a dune environment could begin by examining strategies outlined in the Integrated Coastal Zone Management Plan (ICZM) and the Shoreline Management Plan for the coastline, and identifying specific conservation areas such as SSSIs and National Nature Reserves. As an example, websites providing secondary data on the Sefton dunes in northwest England are given in Table 4.7.

Table 4.7 Websites providing secondary data: Sefton dunes, northwest England

ICZM Plan	www.seftoncoast.org.uk/pdf/iczmplan20062011.pdf
Natural England and Ainsdale NNR	www.naturalengland.org.uk
DEFRA	jncc.defra.gov.uk
Local authorities and universities	www.seftoncoast.org.uk
	www.sandsoftime.hope.ac.uk
Lancashire Wildlife Trust	www.lancswt.org.uk
Shoreline Management Plan	www.seftoncoast.org.uk/pdf/scmp.pdf

Management must balance conservation with public access to the dunes. Possible fieldwork tasks to investigate management strategies are:

- vegetation surveys (quadrat sampling of species diversity, plant cover) comparing least accessible areas in dunes with areas in the vicinity of footpaths, boardwalks, car parks, etc.
- distribution of visitors and identifying areas of acute visitor pressure
- recording the number and distribution of dune blow outs (i.e. dune erosion) and their relationship to visitor pressure
- surveys of litter left by visitors on beaches and in the dunes
- the impact of beach cleaning on embryo and foredunes (note that beach cleaning can often accelerate dune erosion)
- the effectiveness of fencing to encourage dune regeneration, the use of boardwalks through the dunes
- the impact and control of alien plant species such as sea buckthorn

Management strategies for the future

Revised

Modern coastal management views the coast as an integrated and interconnected system. Integrated strategies that aim to achieve sustainable management include **Integrated Coastal Zone Management** (ICZM) and **Shoreline management**.

ICZM

ICZM addresses issues relating to the coastal zone. It brings together stakeholders and decision-makers to allow social, economic and environmental issues to be managed in a sustainable manner. In response to these issues local councils produce ICZM plans to reduce potential conflicts.

Shoreline management

Shoreline management aims to produce sustainable policies for the defence of the shorelines around the UK. These policies, formalised in **Shoreline Management Plans** (SMPs) take account of natural coastal processes, the environment and human needs.

SMPs have been produced for each of the eleven **sediment cells** in England and Wales. They set out a long-term strategy for coastal defence in each cell. As a result, coastal resources and issues such as erosion and flooding are managed together rather than separately.

Two important ideas dominate modern thinking on coastal management:

● human intervention that interferes with natural processes should be minimal

● where intervention is necessary, it should be **sustainable**

SMPs apply one of three possible management options for each stretch of coastline (Table 4.8).

> **Sediment cells** are stretches of coastline where erosion, deposition and movement of sediment occur in isolation from adjacent cells. They are the basic unit of coastal management. Their closed nature calls for careful management to ensure sustainability.

> **Examiner's tip**
>
> Coastal management strategies that result in the loss of land and property to erosion and flooding are often controversial. Case studies should be learned to support balanced and well-reasoned discussion effectively.

Table 4.8 SMP management options

Hold the line	Aims to maintain, and in some cases strengthen, existing coastal defences. This option is justified where the value of properties and infrastructure at risk is greater than the cost of defences.
No active intervention	This option covers most of the coastline of England and Wales. Natural processes operate without human interference.
Managed realignment	Sets back the shoreline and allows the sea to flood areas that were previously protected by hard defences. Managed realignment (a) reduces the costs of maintaining hard defences, (b) allows salt marshes and mudflats, which act as natural defences, to develop and (c) creates new wildlife habitats, i.e. mudflats and salt marshes.

Now test yourself

Tested

5   Outline the three main strategies used in Shoreline Management Plans.

Answer on p. 124

Exam practice

1   Study Figure 4.1 on p. 94.

(a) Describe the movement between stores of sediment in the coastal zone.   [10]

(b) Describe the fieldwork and research you undertook to investigate rates of coastal retreat and their impacts on development and people at a small scale.   [15]

(c) Using examples, explain how coastal management is adapting to new ideas and situations.   [10]

Answers and quick quiz 4 online

Online

5 Unequal spaces

Recognising inequality

Inequality at a variety of scales

Revised

Inequality is about unequal access of different groups to employment, housing, education, healthcare, etc. Groups with least access to jobs, affordable housing and essential services are likely to be poor and suffer **multiple deprivation**. Multiple deprivation **excludes** a significant minority from participating fully in the economic, social, and political life of society.

> **Enquiry question**: What are unequal spaces and what causes them?

Scales and areas of inequality

In England, inequality is found in urban and rural areas, and at inter-urban and intra-urban scales.

- Urban populations are generally more deprived than their rural counterparts. 98% of the most deprived small census areas in England in 2010 were in urban areas (Figure 5.1).

- The most deprived urban areas are conurbations in northern England, particularly Merseyside and Greater Manchester. The most prosperous are in southern England, especially London and smaller towns such as Guildford, Salisbury and Winchester.

- Inequalities are most conspicuous within urban areas. In cities, populations often become spatially **segregated** by income (Figure 5.2).

> **Typical mistake**
>
> Although relative deprivation is much higher in urban than in rural areas in England, absolute levels of inequality and deprivation in rural England are high. In 2010 nearly 1 million rural households lived on less than 60% of the national median income.

The 10 most deprived areas
1 Jaywick Sands, Clacton-on Sea
2 Grant Thorold, Grimsbury
3 Revoe, Blackpool
4 Anfield, Liverpool
5 Grange Park, Blackpool
6 Speke, Liverpool
7 West of Burnley, Lancashire
8 Queenstown, Blackpool
9 Weir, Rochdale
10 Collyhurst, Manchester

Most deprived areas   Least deprived areas

Figure 5.1 England's deprived areas, 2010
(Source: Department for Communities and Local Government)

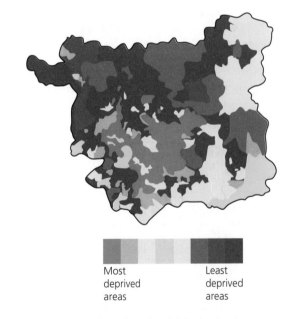

Most deprived areas   Least deprived areas

Figure 5.2 Leeds: index of multiple deprivation

Factors creating levels of inequality

Revised

In the UK inequality is associated with income, housing, employment, education, health, healthcare, crime, and the quality of the living environment.

- Income defines relative poverty in the UK and the EU (i.e. less than 60% of median household income). Low incomes are linked to poor educational attainments and lack of skills and training. Often the result is unemployment or low-wage employment. Employment opportunities may also be limited by inadequate public transport, the cost of childcare, ageism, ill-health and physical disability.

- Access to affordable/adequate housing is closely related to income. Poorer groups are constrained to purchase or rent low-cost housing. Because this housing is concentrated in inner city areas or on peripheral social housing estates, the cost of housing is a major cause of spatial segregation in urban areas.

- In many rural areas attractive to commuters, an increase in the numbers of second-home owners and retirees has resulted in rising house prices, making housing unaffordable to many young local families.

- Educational opportunities may be restricted by low-achieving state schools, disproportionately represented in low-income urban residential areas.

- Chronic ill-health is more prevalent among low-income than high-income groups. Ill-health is related to sub-standard housing, inadequate diet, lifestyle and the stress of living in poverty.

- Accessing healthcare services is difficult for the elderly who often live alone and are less likely to drive or own a car. These problems cause particular hardship in rural areas where GP surgeries and hospitals are widely dispersed.

- Groups suffering multiple deprivation are likely to have greater exposure to crime and low-quality physical environments. High crime rates in deprived neighbourhoods are linked to unemployment, drug use, poor educational attainment and poverty.

Examiner's tip

Remember that inequality has multiple and interrelated dimensions.

Now test yourself

1   List five dimensions of multiple deprivation.

Answer on p. 124

Tested

Fieldwork and research

Revised

Fieldwork and research can explore spatial patterns of inequality in urban and rural contexts. Both urban and rural studies use similar primary and secondary data sources (Table 5.1).

Table 5.1 Data sources for investigating spatial inequality

Access to services	Fieldwork: rural and urban questionnaire surveys on the location of and access to supermarkets, schools, doctors' surgeries, hospitals, dentists, public transport, etc. Respondents grouped according to age and mobility.
Environmental quality and land use	Fieldwork: urban survey to evaluate housing and environmental quality. Location of shops selling fresh food, and negative externalities, e.g. heavy industry, polluted land.
Crime	Local crime maps available at: **www.police.uk/**
Employment	Employment, unemployment, economic activity, benefit claimants at ward levels available at: **www.nomisweb.co.uk/**
Qualifications	Available at ward level: **www.nomisweb.co.uk/**
Multiple deprivation	Multiple deprivation index and its various domains (e.g. income, housing, education, crime) available for wards and LSOAs at: **www.ons.gov.uk/ons/rel/hsq/health-statistics-quarterly/no--53--spring-2012/uk-indices-of-multiple-deprivation.html**
Planning documents	Reports by local authorities on multiple deprivation, with mapping of multiple deprivation index and its domains, and ward profiles. An example (Liverpool) is at: **liverpool.gov.uk/council/key-statistics-and-data/indices-of-deprivation/**

Neighbourhood statistics	Data available for different geographies (e.g. local authorities, wards, parishes, super output areas, output areas) derived from the census and other sources available at the ONS website: **www.neighbourhood.statistics.gov.uk**
	Data sets relevant to inequality include: work deprivation; education, skills, training; air pollution; land use statistics; housing benefit; households in poverty; lone parent families; health and morbidity.

Examiner's tip

Investigations of inequality could be based around contrasting sample areas, e.g. an inner city and outer suburban area, or a pressured (i.e. commuter belt) and remote rural area.

Inequality for whom?

Economic and social exclusion
Revised

Rural areas

Economic and social exclusion among rural populations is closely related to the isolation of rural areas. The key issues concern rural transport, healthcare, employment and housing.

Enquiry question: What impact do unequal spaces have on people?

Rural transport

Low car ownership and inadequate public transport contribute to loneliness, which inhibits the social lives of individuals and families. Poor transport provision also limits access to basic services and employment. Private transport is also more expensive in rural areas. The cost of petrol, for example, is up to 10 p per litre higher. This problem is aggravated because rural dwellers have lower disposable incomes than urban dwellers, rely heavily on private transport, and have comparatively long journeys to work, to shop and meet friends.

Health services

Low-income families have difficulty accessing primary healthcare services. In recent years, the centralisation of medical services has increased travel times and inconvenience for rural patients. Even longer journeys are necessary to access hospitals. Meanwhile there is little rural-based support for job seekers, the homeless, drug and alcohol abusers and women exposed to domestic violence.

Employment

There is a shortage of employment opportunities. Employment in agriculture, once the mainstay of the rural economy, has declined steadily in the past 50 years. Average wages are lower in rural than in urban areas, and employment is more likely to be seasonal and part-time. Poor public transport is a further barrier to finding work in urban areas. Young people, the disabled, women and lone parents are most likely to be unemployed.

Housing

Opportunities for native rural dwellers to buy or rent housing in rural areas have diminished in recent years. This is because in many rural areas:

- a large proportion of the rural housing stock is poor quality and in a poor state of repair
- there is a lack of affordable housing for young families, a situation exacerbated by house price inflation and low wages
- many houses have been sold as second homes and holiday homes
- a large part of the rural social housing stock has been sold under 'right-to-buy' legislation

Lack of affordable housing has forced many young people to leave their native communities and move to towns and cities. This outmigration has contributed to falling school rolls, the closure of small primary schools, and longer journeys to school for children.

Urban areas

Inequality and deprivation cause social and economic exclusion in urban areas. Excluded individuals and groups have fewer opportunities and poorer access to services compared with other citizens.

- Urban areas with the highest levels of deprivation are inner cities and outer peripheral social housing estates. Fears of security and violent crime in these areas may limit the activities and social lives of groups such as the elderly and young women.

- The physical environment of deprived areas is often run-down, housing may be high density, and there are fewer open spaces and limited opportunities for recreation and leisure.

- Air pollution is often highest in the most deprived neighbourhoods. Nitrogen oxide and particulates from vehicle exhausts increase the incidence of respiratory disease.

- Residents in the most deprived areas may experience discrimination from banks, financial agencies and insurers. Motor and household insurance premiums are high in these areas. Some residents may also be credit blacklisted because of their postcodes.

- Banks and building societies may be unwilling to provide mortgages on properties perceived as high risk in deprived areas.

- Local children are likely to attend low-achieving local schools, reducing their prospects of gaining good exam qualifications.

- Low income neighbourhoods are often poorly provided with shops that sell fresh food, including fruit and vegetables. The resulting 'food deserts' and reliance on processed and junk food contribute to health problems such as obesity and diabetes.

- Housing quality is poorest in most deprived areas. Residents may live in cramped conditions, without adequate heating, poor noise insulation and damp. These conditions may cause physical and mental health problems.

Typical mistake

Many people are unaware that deprivation is widespread among rural communities in the UK. Rural deprivation is often underestimated because it is less geographically concentrated than deprivation in towns and cities.

Now test yourself

2 What factors make housing unaffordable to many people who want to remain in rural areas?

3 How does lack of personal mobility affect inequality in rural areas?

Answers on p. 124

Tested

Examiner's tip

Geography (i.e. where people live) is one factor contributing to inequality and deprivation. Access to services and employment are influenced by geography. Meanwhile the location of housing types (as well as income) has a major bearing on where city dwellers live.

Now test yourself

4 What is meant by social and economic exclusion?

Answer on p. 124

Tested

Inequality and marginalised groups

Revised

Inequality creates **marginalised groups** in society, defined by low incomes, poor education, physical or mental disability, ethnicity, age and gender. Discrimination against marginal groups places them in an unequal position compared with the majority.

Marginalised groups in rural areas

In rural areas marginal groups are mainly those with limited access to employment and basic services. The principal cause of rural marginality is immobility. The groups most affected are the elderly, the disabled, the chronically ill, young people searching for employment and one-car families where one parent stays home to care for young children.

Marginalised groups in urban areas

Marginalised groups in urban areas are more diverse. Large ethnic minorities are often marginalised because of poor English language skills and discrimination by the host society. Because of language barriers, culture and tradition, Muslim women born in south Asia are often very isolated, with little contact with mainstream society.

Asylum seekers and illegal immigrants are even more marginalised. The latter have no legal access to social and health services and little prospect of formal employment. Without a permanent address, homeless people are also excluded from the job market. Other groups on the margins of society are ex-prisoners and drug addicts.

Since 2008 youth unemployment has soared in the UK. The result is a large group of disaffected, poorly educated young men, surviving on benefits and unable to share the advantages of consumerism enjoyed by the majority. Exclusion and alienation of this group contributed to the urban riots in London, Manchester, Birmingham and other UK cities in 2011.

Single parent families are also marginalised and are likely to be poorer than nuclear families. Lone parents with young children who cannot afford the cost of childcare are often financially better-off relying on state benefits rather than working. For single parents, low incomes and isolation often mean economic and social exclusion.

Examiner's tip

Remember that equality is about having equal opportunities regardless of social and economic background, ethnicity, gender, age, health, etc.

Fieldwork and research

Revised

Fieldwork and research focuses on criteria to identify spatial patterns of inequality and evaluate schemes designed to tackle inequality (Tables 5.2 and 5.3).

Table 5.2 Criteria for identifying spatial patterns of 'haves' and 'have nots'

Rural areas	
Income deprivation	For Lower Layer Super Output areas (LSOAs — boundaries available at ONS neighbourhood statistics website) see: **www.ons.gov.uk/ons/rel/hsq/health-statistics-quarterly/no--53--spring-2012/uk-indices-of-multiple-deprivation.html**
Housing deprivation	
Unemployment	At ward levels: **www.nomisweb.co.uk/**
Crime levels	For postcode units: **www.police.uk/**
Distance/journey times to services	For supermarkets, doctors' surgeries, dentists, schools: Google Maps.
Service provision	Either field surveys or Google 'Street View' to identify local service provision.
Urban areas	
As above for income, housing, unemployment, crime and distance to services.	
Service provision	Field surveys or Google 'Street View' to locate shops selling fresh food.
Mortality rates	Overall mortality and infant mortality data available at ONS neighbourhood statistics for LSOAs website (see above). This information could be combined with population figures to calculate mortality statistics such as the CDR death rate and IMR.
Segregation	Multiple deprivation index and the distribution of ethnic minority groups (from ONS neighbourhood statistics) could be analysed spatially by calculating the index of dissimilarity (see: **cdu.mimas.ac.uk/materials/unit17/3_the_index_of_dissimilarity.html**

Table 5.3 Checklist to evaluate schemes to tackle inequality

Employment	Jobs created or numbers of unemployed before and after the introduction of employment schemes: **www.nomisweb.co.uk/**
Affordable housing	Fieldwork estimates of the number of new affordable houses completed in a rural area.
Services	The extent to which retail, healthcare, transport and medical services have been maintained or improved. Fieldwork observation and interviews with rural residents.
Education	Changes in the number of students achieving five good GCSE passes at local secondary schools. Comparison with national averages: **www.education.gov.uk/schools/performance/archive/index.shtml**
Crime	Changes in levels of crime: **www.police.uk/**
Segregation	Changes in segregation (dissimilarity indices) using 2001 and 2011 census data.
Deprivation	Comparison of 2004, 2007 and 2010 multiple deprivation indices for LSOAs: **www.communities.gov.uk/communities/research/indices deprivation/deprivation10/**

Managing rural inequalities

Rural inequalities

In rural areas the key problems that contribute to inequality are lack of employment opportunities, shortages of affordable housing, lack of basic services and poor infrastructure.

Local employment

Barriers to investment and employment in rural areas are:

- shortages of work premises for small businesses
- absent or slow internet connections
- limited training opportunities
- poor infrastructure

Affordable housing

Lack of affordable housing remains a serious issue in the countryside, especially in many coastal areas, National Parks and Areas of Outstanding Natural Beauty (AONB). Planning controls restrict new building in these areas. Because of the limited housing stock, growing demand and low wages, local people are priced out of the market. Without affordable housing local people move away and rural services close.

Healthcare

Healthcare problems centre on (a) providing effective primary and secondary healthcare to scattered and remote rural populations, and (b) patients' difficulties in accessing healthcare services where public transport is limited and where poorer rural households cannot afford a car.

Cost-effectiveness and increased medical specialism has led to the centralisation of healthcare services. Many small community hospitals have closed and single GP surgeries have been centralised in modern medical centres in small towns. These trends have disadvantaged rural communities who have to undertake longer journeys.

Retailing and transport

The sustainability of shops and transport services in rural areas depends primarily on the level of demand.

Retailing

Rural village stores, pubs and post offices continue to close because:

- many residents (e.g. commuters) are mobile and prefer to shop in large supermarkets in nearby towns where goods are cheaper and there is more choice
- the development of large supermarkets that service rural as well as urban hinterlands
- outmigration of rural people undermines the thresholds for sustaining local retail services

Transport and communications

Rural bus services rely heavily on subsidies from local authorities. But with the squeeze on public spending in the UK since 2008 many services have been withdrawn. Despite local authorities spending around £800 million per year subsidising rural buses, one in ten rural communities has no bus service.

The absence of broadband in many rural areas is a barrier to business, investment and employment. It also excludes shoppers from the advantages and convenience of e-commerce. The government is committed to providing all rural areas with a minimum of 2 Mbps broadband speed by 2015.

> **Enquiry question**: How can we manage rural inequality and improve the lives of the rural poor? How successful have particular schemes been?

> **Examiner's tip**
>
> Sustainable rural communities are only possible when basic services and affordable housing are available. The decline of basic services and affordable housing create serious economic and social barriers that contribute to inequality.

Exam practice answers and quick quizzes at **www.hodderplus.co.uk/myrevisionnotes**

Fieldwork and research

Primary and secondary sources can be used to investigate the effectiveness of policies to reduce rural inequalities.

Primary sources

Questionnaire surveys of rural residents could aim to investigate:

- recent community-based schemes to improve access to transport services (e.g. voluntary car schemes, taxi-sharing)
- voluntary schemes to create sustainable retailing services (e.g. cooperatives)
- changes in local employment opportunities and the local availability of primary healthcare services
- the construction of affordable homes
- internet availability, its speed and the impact on residents' lives

People's views on new developments and their perception of the extent to which they have reduced inequality could be explored. Questionnaire surveys could be completed by field observation of new buildings, the use of services such as village shops and the opening hours of doctors' surgeries.

Secondary sources

Look at websites for information on (a) projects that affect the area studied and (b) general policy initiatives, statistics and trends in rural England. Possible websites are the local authority responsible for the rural area (education, planning, infrastructure), housing associations, DEFRA (initiatives aimed at rural development), the local Primary Care Trust, Department for Communities and Local Government (for general rural policy initiatives), and Action with Communities in Rural England (ACRE).

Typical mistake

Poverty in the UK is not confined to towns and cities. Around 3.5 million people in rural districts live in households with low incomes. In rural areas the two most important barriers to equality are access to housing and transport.

Managing urban inequalities

Urban inequalities

Revised

Social problems

The most urgent social problems in deprived urban areas are crime, drug and alcohol abuse, anti-social behaviour and a culture of worklessness.

In residential areas, urban crime rates are highest in the inner city and on social housing estates. Although drug and alcohol abuse are widespread across society, they are more prevalent in areas of low income and deprivation.

Anti-social behaviour occurs where gangs and youths assemble on the streets, and where families have little respect for their neighbours. These problems are most prevalent on estates of social housing and are linked to issues of multiple deprivation.

Many families have no working adults, a situation that may extend back several generations. Long-term worklessness creates a mentality of dependence, a low work ethnic, and a reluctance to go out and find employment. Worklessness is also a function of poor qualifications and skills.

Enquiry question: What strategies can be used to combat inequality in urban areas? How successful have particular schemes been?

Economic problems

Poverty is most often caused either by low wages or unemployment. In the poorest areas of large cities, typically:

- 40% of families have incomes that are 60% below the national median income (official definition of poverty)

- only 40% to 50% are economically active
- between 40% and 50% of adults have no qualifications or skills

Environmental problems

Air pollution problems are often worst in the most deprived urban neighbourhoods. Environmental inequalities arise because the poorest members of society experience the highest levels of pollution. The principal pollutants are NO_2 and Pm_{10} from vehicle exhausts, and SO_2 from power stations and heavy industry. Air pollution increases rates of respiratory disease such as asthma, bronchitis and lung cancer.

Poor-quality built environments are often the backdrop to deprived urban neighbourhoods. In cities such as Glasgow and Birmingham blocks of high-rise flats survive from the 1960s. For decades they have been unpopular with tenants and are unsuitable for the elderly and young families. Inner city districts with high residential densities have few green and public spaces. Unused land often raises issues of visual pollution and community safety. Poor street lighting adds to fears of crime.

Key players involved in delivering solutions

Urban social, economic and environment problems are often highly localised. This geographical concentration has advantages: it allows resources to be focused on specific areas, and the decisions and actions of agencies and key players to be coordinated. Given the interrelated nature of urban problems, an area-based approach makes a lot of sense.

In the UK, **urban regeneration**, through **Area Action Plans**, has been the main response to the problems of multiple deprivation. Current high-profile examples are the East Leeds and Southeast Leeds (EASEL) and the New East Manchester (NEM) regeneration projects. These **partnership** projects between the public and private sectors also involve the voluntary sector and local communities. They aim to achieve social and economic, as well as physical regeneration. The **key players** are:

- central government, through Housing Action Plans, with the Safer Stronger Community Fund providing part of the funding
- local councils
- developers and building contractors
- housing associations, which may buy and let new and refurbished affordable housing
- housing management companies, maintaining and administering houses owned by the council
- companies involved in skills and retraining programmes for local residents
- voluntary sector organisations, e.g. charities setting up sports clubs, wildlife and conservation projects
- the local community, through consultation meetings with councillors, planners and developers, contributing to decision-making

> **Examiner's tip**
> You should note that most urban regeneration schemes are housing or property-led.

> **Key players** are the main decision-makers in regeneration schemes. They include planners, local councillors, developers, property agents, charities and, increasingly, local residents.

> **Now test yourself**
> 5 Name four key players in urban regeneration schemes.
> **Answer on p. 124**
> Tested

Fieldwork and research

Revised

Fieldwork and research focus on ways to reduce urban inequalities such as self-help, traffic and public transport schemes, planning initiatives, business initiatives and crime and policing policies.

Self help

- Shanty builds on garden plots in response to housing shortages, especially for immigrants. Investigate this problem in areas such as Southall in west

London. Locate the shanties on Google Earth, and support with field surveys. Conduct a questionnaire survey among local residents and tenants. Evaluate the impact of shanties and the pros and cons of a self-help solution more typical of cities in LEDCs.

- Residents' associations are concerned with matters of local interest such as planning, crime, amenities and charitable work. They are a powerful voice representing local opinion. Their impact in reducing inequality could be investigated qualitatively through interviews with committee members and other residents. Comparison with an area lacking a strong residents' association might be revealing.

- Neighbourhood Watch is a voluntary partnership between the local community and the police, aiming to prevent or reduce local crime. Exposure to high crime rates is a feature of urban inequality. The effectiveness of Neighbourhood Watch could be investigated by (a) researching crime maps for postcode units, (b) interviewing local residents on their perceptions of security, and (c) comparing crime rates in similar communities with and without Neighbourhood Watch schemes.

Traffic and public transport

The cost of using urban public transport contributes to inequality. A number of initiatives by local transport authorities should help reduce inequality:

- *Transport for London*'s Oyster cards include concessions for children, students, benefit claimants, the disabled and the elderly, which make transport by the tube, buses and trains more affordable
- local authorities provide travel cards, making bus transport free for the over 60s
- railway operators provide discount travel through rail cards for students, senior citizens and others

Congestion and air pollution problems often hit poorer inner-city communities hardest. Schemes to reduce the flow of private vehicles and air pollution in towns and cities include 'park-and-ride' and, in London, the congestion charge and low emissions zone.

The effectiveness of these and other schemes could be assessed from secondary data on usage, 'before and after' traffic flows and pollution counts, and interviews with residents who (a) benefit from public transport concessions and (b) live in the most polluted and congested neighbourhoods.

Planning initiatives

Major planning initiatives over the past decade centre on Area Action Plans (AAPs). They address inequality in the most deprived areas through partnerships between local councils and the private sector. Although largely property-led, they also target unemployment, education and skills training, crime and so on.

- Select an Area Action Plan that targets areas of multiple deprivation in a nearby town or city. Research the AAP through the internet.
- Through fieldwork, assess the quality of affordable housing and the residential environment in areas of new-build.
- Compare the new-build areas with areas of older housing undergoing refurbishment.
- Interview local residents to obtain their views on regeneration, their awareness of the AAP, and the extent to which they have been consulted about development.
- Investigate the perceptions of local residents concerning the impact of regeneration on (a) crime, (b) education and skills training, (c) employment and (d) service provision.

Business initiatives

- Some of the UK's leading retailers, such as Tesco and Asda, have opened supermarkets in deprived urban areas. In partnership with local councils, they employ and train local people to work in the stores. The success of such schemes (local recruitment, training, retention, promotion to better paid jobs, etc.) could be investigated by interviewing store managers and local employees.

- Partnership schemes encourage enterprise and help unemployed people in deprived areas to set up their own businesses. Investigations could look at new businesses, the type of support provided (advice, grants, premises), employment created and the businesses' achievements and problems.

- Furniture recycling schemes provide affordable furniture, donated by the public, to disadvantaged groups such as benefit recipients, people on low incomes, OAPs and students. Research the schemes operating in an area. Interviews with key players in the charities involved could provide insights into supply and demand for recycled furniture, the contribution of the service to reducing inequality and how the service could be made more effective.

Examiner's tip

The success of schemes to reduce urban inequalities could be measured (a) objectively, using information released by local authorities, the Environment Agency, the police and other public bodies or (b) subjectively, by investigating the perceptions of local people of the schemes' effectiveness.

Crime and policing

Video surveillance and neighbourhood policing could be introduced to reduce crime, especially in the most deprived residential areas where crime levels are highest. Possible questions for investigation are:

- Do people feel safer in areas (a) monitored by video cameras, (b) with neighbourhood policing?

- To what extent are people aware of (a) video surveillance, (b) neighbourhood policing?

- Have crime rates fallen in areas subject to (a) video surveillance, (b) neighbourhood policing?

Exam practice

1 Study Figure 5.2 on p. 103.

   (a) Comment on the spatial distribution of multiple deprivation in Leeds. [10]

   (b) Describe the fieldwork and research you undertook to investigate inequality in either an urban or a rural area. [15]

   (c) Using examples, assess the effectiveness of efforts to manage inequality in either urban or rural areas. [10]

Answers and quick quiz 5 online

Online

6 Re-branding places

Time to re-brand

How places re-invent and market themselves

Revised

What is re-branding?

Re-branding is about marketing a place to give it a new identity in the consciousness of the public and business. It may appeal to local, national and international markets (Figure 6.1).

Re-imaging contributes to re-branding by countering negative images of places (e.g. industrial decline, deprivation, crime) with booster imagery that promotes modernity, enterprise and post-industrial functions (e.g. leisure, tourism). It is essentially a public relations exercise. Re-imaging emphasises the individuality of a place and the things that makes it different or 'special', and usually includes a designer logo (which will appear on all publications) and a slogan. Thus Norwich markets itself as 'a fine city'; Liverpool is 'the world in one city'; and Yorkshire is 'alive with opportunities'.

Regeneration is a long-term process involving economic, social and physical actions. Along with re-imaging it contributes to re-branding by reversing decline and creating sustainable communities.

> **Enquiry question**: What is re-branding and why is it needed in some places?

Re-imaging
Re-imaging contributes to re-branding, creating new mental pictures of places in the minds of consumers

Re-branding
Re-branding is marketing a place to give it a new identity in the consciousness of the public and business. It may appeal to local, national and international markets

Regeneration
Regeneration is a long-term process involving economic, social and physical improvment of a place

Figure 6.1 The process of re-branding

Examiner's tip

Image plays a vital role in determining human behaviour. People act and make decisions according to their perceptions of the world, rather than objective reality. PR and advertising can therefore make a significant difference to places and their attractiveness to investors and visitors.

Now test yourself

1 What is meant by re-branding and re-imaging?

Answer on p. 124

Tested

Case study Ideas for re-branding places

Liverpool: Culture as a catalyst. Liverpool was the European Capital of Culture (ECoC) in 2008. Its year-long status as ECoC was marketed as *Liverpool08*. The city was awarded a budget of £130 million to stage more than 275 cultural events. Political and commercial interests in the city saw Liverpool's ECoC status as an opportunity to transform the city's national and international image, and drive regeneration. On the back of EcoC, public and private partnerships were expected to invest around £4 billion in Liverpool between 2008 and 2016.

An official report on the impact of Liverpool's year as ECoC was very positive. The city was described as having undergone an 'image renaissance', locally, nationally and internationally. Levels of confidence were raised across the city, and Liverpool had been rediscovered as a tourist destination beyond its two Premier League football clubs and The Beatles.

Wessex: Heritage and landscape. The Wessex Tourist Board has created an effective heritage brand to promote tourism in Dorset, Wiltshire and Somerset. Its website describes Wessex with the slogans 'a world of heritage at your feet' and 'the heart of ancient England'. Prehistoric monuments from the Neolithic to the Iron Age, Saxon history and Alfred the Great are mixed with the myths and legends of Glastonbury and King Arthur. The region was also the setting for Thomas Hardy's nineteenth-century Wessex novels, and tourists can follow Hardy trails and visit places referred to by pseudonyms in the novels such as Casterbridge (Dorchester) and Sherton Abbas (Sherborne). They also visit locations used in film and television adaptations of Hardy's novels.

Wessex's historical and literary associations are set against a landscape that offers outstanding recreation and leisure opportunities, which includes national parks (New Forest, Exmoor), Areas of Outstanding Natural Beauty (Cranborne Chase) and a World Heritage Site (the Jurassic Coast).

Notting Hill: Market-led change. Fifty years ago Notting Hill was one of the most run-down parts of west London. It was the scene of race riots in 1958 and, in 1976, following the Carnival, riots resulted in 160 injuries and 66 arrests. In the 1950s and 1960s Notting Hill was home to large numbers of Afro-Caribbean immigrants (the annual Notting Hill Carnival dates from this period). Housing was multi-occupancy and landlords charged extortionate rents. As late as the 1970s, Notting Hill was described as 'a massive slum ... crawling with rats and rubbish'.

Today, Notting Hill presents a different image. It has become one the most fashionable districts in central London. The up-market image of Notting Hill was encapsulated in the 1999 film of the same name, starring Julia Roberts and Hugh Grant. Private investment though **gentrification** has regenerated and re-branded Notting Hill. Large Georgian and Victorian houses, once multi-occupied, have either been converted to desirable flats or returned to town houses. Average house prices are close to £1 million. Most residents are high-income professionals working in producer services, media and design. Meanwhile, shops, bars and restaurants have expanded to serve wealthy residents.

Why re-branding is needed

Revised

Re-branding of places may be needed:

- to increase public or consumer awareness of changes that have/will improve social, economic and physical characteristics of a place
- to emphasise the distinctiveness and attractiveness of a place
- to discard negative images associated with deindustrialisation, public disorder, crime, multiple deprivation, etc.
- to promote pride in a place by local communities
- because places must compete to attract inward investment and visitors

Liverpool: Re-branding to counter urban decline

Liverpool and the surrounding Mersey region form a conurbation with a population of around 1.5 million. The city grew rapidly in the eighteenth and nineteenth centuries, largely as a result of the city's status as a major port benefiting from trans-Atlantic trade, including the slave trade. By 1881 Liverpool's population had reached 500,000.

But as the twentieth century progressed Liverpool entered a period of economic and social decline.

- Between 1930 and 2001 Liverpool lost almost half of its population.
- By the late 1970s and early 1980s the city had lost most of its port trade and the old dockland areas lay derelict.
- Unemployment was high and the workforce of low skill.

In the national consciousness Liverpool had a poor image with a reputation for militancy (among workers and politicians) and 'a declining way of life'. Inequality and poverty sparked riots in Toxteth (one of the city's poorest neighbourhoods) in 1981, an event which reinforced Liverpool's negative image. In 1994 Merseyside qualified for EU Objective 1 status as one of the poorest regions in the EU, with a GDP 75% below the EU average. But despite progress, even in 2010, one-third of the most deprived LSOAs in England were on Merseyside.

> **Examiner's tip**
>
> Give some critical thought to the idea (assumed in re-branding strategies) that places can be treated like products sold in a market place.

> **Now test yourself**
>
> 2 Describe three types of places that might adopt re-branding strategies.
>
> **Answer on p. 124**
>
> Tested

Fieldwork and research

An investigation into the profile of places needing re-branding, using primary and secondary sources, might consider:

- environmental quality: fieldwork surveys of housing quality, the extent and distribution of unused or derelict land, secondary sources such as Google Earth, Street View and photo panoramas
- retailing: fieldwork surveys of the quality of shops in the CBD, the number of vacant retail units, evidence of recent investment, retail decline (comparison of survey results with recent Goad maps)
- census data at ward and middle or lower level super output areas of employment, incomes, benefit recipients compared with national averages
- multiple deprivation data (2010) at ward or lower level super output areas placed in rank order to assess deprivation in a national context
- age-related mortality data, e.g. life expectancy, infant mortality, standardised mortality rates at ONS website: **http://www.neighbourhood.statistics.gov.uk**
- public and private sector employment

Re-branding strategies

Stakeholders in the re-branding process

Re-branding is often a partnership between the public and private sectors. The key players are **stakeholders** who have an interest in the project. They may include local authorities, government agencies, investors, property developers, real estate agents, land owners, PR and advertising agencies, local communities, visitors and others. The interests of some principal stakeholders are:

- local authorities' and government agencies' interests may be to stimulate economic growth, create new employment, improve the physical environment (e.g. by reclaiming **brownfield sites**), improve local housing, etc.
- property developers will liaise with investors, land owners and building contractors: their interest is profit-making
- real estate agents are responsible for letting commercial and residential properties. Like developers, their primary interest is profit
- local communities, who are consulted in most regeneration schemes, will be most concerned about the social, economic and environmental benefits to them, i.e. new employment opportunities, new housing and new amenities

Re-branding strategies

The main re-branding strategies are market-led, top-down, flagship projects, legacy and catalysts.

- Market-led re-branding involves private companies and business people aiming to make a profit from investment. The gentrification of central London districts such as Notting Hill, Islington and Clapham is an example of this approach.
- Top-down re-branding involves large organisations such as councils, planning departments, development agencies and companies in the private sector. Controlled from the top, it is coordinated and well planned. Examples include dockland developments in the 1980s and 1990s such as Salford Quays and Cardiff Bay.
- Flagship developments are large-scale, one-off property projects that act as a catalyst to attract further investment and regeneration. Typical flagship

> **Enquiry question**: Who are the 're-branding players' and what strategies exist for places to improve themselves?

> **Examiner's tip**
> Re-branding can occur as an incidental outcome of regeneration (e.g. gentrification) or as a carefully planned strategy, where a desired new brand is defined, and regeneration works to support and fulfil the brand.

developments include convention centres, sports stadia, luxury shopping malls, museums, etc.

● Legacies, following international sporting events such as the Commonwealth Games in Manchester in 2002 and the London Olympics 2012. Both events helped to re-brand formerly run-down areas in east Manchester and Stratford, respectively.

● Catalysts provided by events or themes designed to promote change, e.g. European Capital of Culture.

Now test yourself

3 Who are the main stakeholders in re-branding projects?

4 What are: flagship developments, market-led re-branding, and top-down re-branding?

Answers on p. 124

Tested

Fieldwork and research

Revised

Rural strategies

Rural re-branding involves a place re-inventing itself to achieve a more prosperous future. Fieldwork and research to investigate rural re-branding strategies might cover the promotion of tourism, adding local value, expanding rural technology, and rural diversification. It is important to identify the main stakeholders and the media used to publicise the strategies.

● Rural tourism: research the promotion of heritage sites (e.g. castles, abbeys, museums, heritage railways), places with strong literary/artistic associations (e.g. Brontë country, Constable country), film locations for popular TV series (e.g. Holmfirth and 'Last of the Summer Wine' country), stately homes (e.g. Chatsworth, Woburn). Are guided tours available? How have local businesses responded? What facilities are available for visitors? Search tourism leaflets and magazines produced by local tourist boards.

● Adding local value: research the efforts of local farmers to supply local markets with home-grown food and organic produce, engage in farmers' markets, support local food festivals (e.g. Ludlow), and operate farm shops, etc. Search local newspapers advertising farm produce and events.

● Rural technology: research the availability and speed of rural broadband and community radio. Investigate the role of telecommunications in attracting quaternary-sector workers to live in rural areas.

● Rural diversification: investigate the re-use of old farm buildings for business offices, residence and tourism, and the diversification of farmers into B&B, caravan and camping sites, etc.

Urban strategies

The image of a large urban area is often defined by its CBD. For most non-residents, a city is its CBD. Urban re-branding strategies therefore often place strong emphasis on regenerating the commercial heart of towns and cities. Fieldwork can be conducted in the CBD to establish its role in re-branding. Evidence of regeneration and change that supports the re-brand might include:

● town centre management (e.g. street cleanliness, street furniture, shop frontages, pedestrianisation, public art)

● new shopping opportunities (e.g. new investments in malls)

● the changing importance of service retailing, with more restaurants, bars, clubs, coffee houses

● opening and closing times of service retailers — the concept of the '24-hour city' (pedestrian counts?)

● the development of cultural quarters and specialised areas (e.g. high-class retailing, markets, galleries, arts)

● new flagship developments for the arts (e.g. concert halls, arenas, museums), investments by major retailers, restoration of historic buildings

● the quality and frequency of cultural events (e.g. rock and orchestral concerts, drama, ballet)

- the extent of video surveillance to improve visitor security
- the development of waterfront sites and quality apartment building in and on the outskirts of the city centre
- improvements in transport — tram systems, more frequent services, less polluting buses, renovated train and bus stations, cycle lanes/cycle storage/cycle hire
- the creation of green spaces — parks and piazzas

Secondary information to support fieldwork observation could be sourced from websites operated by local authority planning departments, developers, local arts bodies, Goad maps, commercial directories, local transport providers, etc.

Re-branding for a sustainable future   Revised

Barcelona is an outstanding example of successful re-branding. Since the early 1990s it has been transformed from a tired industrial port city to a stylish modern city, with media, design, marketing and other service activities and a global image. Using Antoni Gaudi's modernism architecture as a selling point it has also become an international tourism centre.

Barcelona used two high-profile international events — the 1992 Olympic Games and 2004 Universal Forum of Cultures — as catalysts to regenerate some of the city's most run-down areas and re-brand.

The Olympic Games' physical legacy — sports facilities, housing, hotels and improved transport — have created long-term and sustainable benefits. Regeneration has transformed the waterfront, once blighted by old manufacturing industries, into a major tourist district. This area includes the Olympic Village and harbour, new beaches, a riverside park, business area, international conference centre, and media park. The old inner-city industrial area of Poblenou has also been redeveloped. Regeneration has incorporated city planning and new infrastructure (including high-speed rail links to Madrid and France) that meet many sustainable social, environmental and economic criteria.

> **Examiner's tip**
>
> Think of sustainability in the context of re-branding as being about the long-term and permanent benefits of legacies (physical, economic, social, psychological) from major international events such as an Olympic Games or a G20 summit.

The London 2012 Olympics are likely to have similar positive and sustainable impacts on east London. Physical regeneration has provided a sustainable legacy that includes the Olympic village, housing, sports stadia, park and environmental upgrades. This Olympic legacy, together with its huge shopping mall and international transport links, have re-branded the district as an extension of central London.

Managing rural re-branding

Success of re-branding   Revised

The success of re-branding rural and urban places can be measured against a number of criteria. Some are listed below.

> **Enquiry question**: How successful has re-branding been in the countryside?

- Has the mental image of a place in the minds of investors, residents, visitors and consumers changed?
- Have investment by businesses and spending by visitors and consumers increased?
- Have local traders experienced an increase in their turnover and profits?
- Has unemployment fallen?
- Is housing more available and more affordable?

● Have employment opportunities and access to and quality of services improved?

● Have crime rates fallen?

● Has there been an improvement in the quality of the built environment and the natural environment (less pollution, increase in wildlife and biodiversity)?

Fieldwork and research

Rural tourism

Re-branding, supported by regeneration projects, is re-positioning many British coastal resorts upmarket. In northwest England, flagship projects include the refurbished pier at Southport, and Morecambe's renovated Art Deco Midland Hotel. In Cornwall, **destination tourism** promotes food and restaurants at Padstow, art at St Ives, surfing at Newquay.

The success of rural re-branding can be investigated using comparative studies (re-branded and non-re-branded places). Primary data can be obtained from:

● observation and recording of retail units, including the proportion of vacant units

● the types and quality of retailing

● footfall along promenades, piers, harbours and main shopping areas

● questionnaires and street interviews of visitors to determine (a) their perceptions of re-branding, (b) reasons for their visit, (c) places of origin and (d) frequency and type of visit (tourist, day visitor)

● questionnaire surveys of residents to investigate perceptions of change and conflicts that might have arisen (e.g. traffic congestion, noise)

Secondary data that could be used to assess the impact of regeneration and re-branding might include comparison of 2004 and 2010 multiple deprivation data, unemployment data (looking at seasonal unemployment), Goad maps in the context of retail change, promotional materials in the media publicising re-branding.

> **Examiner's tip**
>
> Street interviews are the simplest way to source questionnaire data. Respondents should be chosen either randomly or systematically, questionnaires should be brief, and the questions should be clear and unambiguous.

Rural diversification and the post-productive countryside

In the UK the **post-productive** countryside is one with lower-intensity cropping and livestock production, and fewer agro-chemical inputs. Since the mid-1980s, changes to the EU's Common Agricultural Policy (CAP) have placed more emphasis on the sustainability of farming and its environmental impact, and rural diversification and strengthening of rural communities.

> **Post-productivism** describes a shift in agriculture since the 1980s from intensive to less intensive food production owing to concerns over food quality and environmental impact of agriculture.

Within a defined rural area, investigations into the success of rural diversification could look at:

● non-agricultural farm enterprises, their growth, contribution to farm budgets, re-use of redundant farm buildings (e.g. B&B, farm visits, rented office space, nurseries), wind turbines

● attempts by farmers to market produce direct to consumers (rather than through supermarkets) to increase profits and reduce 'food miles' (e.g. farmers' markets, farm shops)

● farmers converting to organic production of fruit, vegetables, dairy and meat products

● the importance of environmental subsidies (e.g. environmental stewardship) to farm budgets, the impact on the countryside (e.g. woodland, habitat conservation) and wildlife (e.g. bird populations)

● the increasing popularity of the countryside for outdoor recreational pursuits such as rambling and bird watching

Primary data could be collected through interviews with farmers, rural shopkeepers and rural residents, and observation (e.g. changing use of farm buildings). Information concerning grants for **cross compliance** with environmental schemes and for rural diversification projects is available from DEFRA.

> **Cross compliance** means that farmers receive subsidies from the CAP in return for delivering environmental benefits, e.g. leaving uncultivated areas for wildlife, limiting the use of agro-chemicals.

Managing urban re-branding

Fieldwork and research — Revised

Fieldwork and research should be undertaken to investigate the success of specific urban re-branding schemes.

> **Enquiry question:** How successful have urban areas been in re-branding themselves?

Along with a marina and the Museum Quarter, The Deep — a modern aquarium — is the centrepiece of Hull's regenerated docklands and waterfront. The success of the re-branding can be assessed by:

- comparing the present-day physical environment of the docklands and waterfront area with photos of the area prior to regeneration
- sourcing statistical data on visitation to The Deep
- interviewing local retailers on the difference redevelopment has made to trade
- interviewing visitors to investigate their spending, origin, other attractions visited on their trip, and their views on the success or otherwise of re-branding
- measuring footfall at various locations on the waterfront

History and culture

History and culture are the main visitor attractions of cities such as York, Edinburgh and Bath. Historic images and brands could be investigated by researching old railway and tourism posters. These could be compared with recent advertising images to determine the direction of re-branding. The impact of re-branding on the mental images that visitors have on arrival in the city (e.g. landmark buildings, physical features, cultural features) could be investigated by interviews. The success of re-branding could also be measured using methods similar to those for flagship schemes.

Sport and leisure provision

Major international sporting events, such as the 2002 Commonwealth Games in Manchester, and the 2012 London Olympics, have been used as instruments for re-branding urban areas. Both events were focused on run-down inner-city neighbourhoods, badly in need of regeneration. The extent to which re-branding has succeeded can be measured by:

- observing and mapping through fieldwork, aerial photos and Google Earth, land use changes
- assessing the quality of the physical environment in regenerated areas and comparing it with neighbouring areas that did not figure in the event
- sourcing statistical data from local authorities on the provision of new and affordable housing and the number of permanent jobs created for local residents
- identifying any 'catalyst effect' on businesses, i.e. the stimulus given by new stadia and other sports facilities to investment in the area
- analysing the socio-economic impact on the immediate area and neighbouring areas using census data (e.g. small areas' statistics on

> **Examiner's tip**
>
> When devising questionnaire surveys you need to decide whether the information you want is quantitative or qualitative. Often, when investigating people's attitude, opinions and mental images, descriptive and qualititative responses provide more detailed and more useful information.

employment, benefit claimants, crime) and data on multiple deprivation (compare 2004 with 2010 figures) from ONS

- investigating through interviews, the 'before and after' mental images of the event area of local residents and people living elsewhere
- investigating through interviews, the opinions of local residents on the overall impact of the event

Now test yourself

5   What criteria could be used to measure the success of re-branding in a coastal resort?

Answer on p. 124

Tested

Exam practice

1   Study Figure 6.1 on p. 113.

   (a)   Comment on the contribution of re-imaging and regeneration in the re-branding of places.   [10]

   (b)   Describe the fieldwork and research you undertook to investigate the strategies used to re-brand either rural or urban places.   [15]

   (c)   Using examples, explain how successful either rural or urban areas have been in re-branding themselves.   [10]

Answers and quick quiz 6 online

Online

Now test yourself answers

Chapter 1 The world at risk

1 A natural hazard becomes a natural disaster when it has a severe impact on human populations.

2 Earthquakes and cyclones are naturally occurring physical events. They become hazards when they adversely affect people.

3 Vulnerability to hazards is influenced by poverty and ability to cope, population density, and the preparedness of society.

4 Disaster risk is directly related to the size and scale of natural hazards and a population's vulnerability, and inversely related to its ability to cope with disasters.

5 Flood disasters appear to be increasing because of (a) climate change and more extreme rainfall, and (b) population growth increasing exposure to river and coastal flood risks.

6 Global warming (a) increases sea surface temperatures (SSTs) in the tropics and sub-tropics, which provide the energy to drive tropical cyclones, (b) increases evaporation and allows the atmosphere to hold more moisture, and (c) changes the position of the jet stream, which controls the movement of depressions.

7 El Niño is a cyclical temperature anomaly that warms the surface waters of the eastern Pacific Ocean. Part of ENSO, El Niño causes extreme weather conditions in the Pacific Basin and beyond.

8 The number of deaths caused by natural disasters (relative to population size) is decreasing because of a decline in vulnerability. Societies are better prepared to mitigate, and cope with, natural disasters.

9 The economic cost of natural disasters is increasing because of higher levels of fixed investment in buildings and infrastructure. The value of fixed assets (covered by insurance) has risen with global urbanisation.

10 Three natural hazards associated with tropical cyclones are hurricane-force winds, sea surges and torrential rain.

11 The earthquake focus is the exact location within the Earth's crust where rocks fracture and release powerful earthquake waves. The epicentre is the location on the Earth's surface directly above the focus.

12 Landslides often occur in association with cyclones, earthquakes and volcanic eruptions.

13 Africa is the continent most affected by drought.

14 A 'disaster hotspot' is a place that experiences multiple natural hazards.

15 Glacials are cold climatic periods (ice ages) lasting tens of thousands of years. Inter-glacials are shorter spells of warmer climate when ice sheets and glaciers retreat.

16 The landscape evidence of climate change in Britain's uplands includes landforms caused by the erosion and deposition by ice sheets, glaciers and meltwater. In addition there are periglacial landforms (e.g. screes, stone circles) developed by frost action and ground ice during tundra-like climate conditions.

17 The so-called 'Little Ice Age' describes the cooling of the Earth's climate between the mid-fourteenth and early nineteenth centuries. The result was winter temperatures in the UK that were 1° or 2°C lower than today's.

18 Astronomical causes of climate change include changes in the Earth's orbit, changes in the angle of the Earth's axis, and the precession of the seasons.

19 The greenhouse effect is the natural warming influence of gases such as CO_2, CH_4 and water vapour (greenhouse gases or GHGs) on the atmosphere. Warming has increased rapidly in the past 200 years due to human activity and GHG emissions. This is known as the enhanced greenhouse effect.

20 Rapid warming in the Arctic is due to changes in the region's albedo. As ice (which is highly reflective) melts it exposes land and ocean surfaces that absorb more of the Sun's radiation. The result is a rapid rise in average temperatures.

21 Global warming in Africa will mean longer and more intense droughts and increased flooding. Large parts of Africa are semi-arid and small reductions in rainfall in these areas could make farming unsustainable. Africa is also the poorest continent. The number of people living in poverty in Sub-Saharan Africa increased in the past decade. These people do not have the resources to cope with climate change.

22 (a) A worldwide rise in sea level is a 'eustatic' rise. (b) Sea level is currently rising worldwide because of (1) melting of ice sheets and glaciers, and (2) the expansion of the oceans as surface water temperatures increase. Both processes are linked to global warming.

23 Predicted sea level rises of 1–2 m in the South Pacific during the next 50–100 years will swamp some coral atolls making them uninhabitable. Evacuation from islands at greatest risk is already occurring. Elsewhere coastal settlements under threat will be re-located on higher ground. Sea level rise will also damage coral ecosystems (major food sources for islanders). Food production is also threatened by salt water intrusion of groundwater.

24 (a) Tipping points are critical thresholds in natural systems that, once exceeded, trigger abrupt and irreversible change. (b) Positive feedback occurs in a system when change induces further change and instability. Rapid melting of ice in the Arctic is an example of positive feedback.

25 Governments can reduce GHG emissions by (1) promoting the use of renewable energy, (2) imposing green taxes, and (3) promoting energy conservation.

26 Cities can adapt to warmer climates in future by planting drought-resistant tree species, planting more trees in streets and parks, planting vegetation on rooftops, installing air conditioning in public buildings, and using permeable building materials to reduce runoff, and conserve soil moisture and urban water supplies.

27 Sustainable adaptations to reduce future threats of flooding and coastal erosion include re-afforesting upland

catchments, controlling urban development on floodplains, conserving wetlands, encouraging the expansion of salt marshes and mudflats.

28 China, India and other developing countries were exempt from the Kyoto Treaty because (1) controls on their GHG emissions would hamper their economic development, and (2) the problem of global warming was largely caused by industrialisation of MEDCs and therefore, morally, their responsibility.

29 IPCC = Intergovernmental Panel on Climate Change, NGO = Non-governmental organisation, UN = United Nations, UNEP = United Nations Environment Programme, GHG = greenhouse gases.

30 The risks of natural hazards are increasing because of (1) climate change, (2) population growth, (3) land degradation, (4) greater fixed investment in urban areas in infrastructure, housing, businesses, etc.

31 Food insecurity exists when people's access to sufficient food needed for a healthy life is no longer assured. Famine is a severe food shortage that leads to a sharp rise in regional mortality. Malnutrition is a lack of adequate nutrition caused by an unbalanced diet. Undernutrition is insufficient food intake to maintain health and well-being.

32 Carbon free energy is an energy source that does not emit CO_2 and other GHGs (e.g. solar energy). Carbon neutral energy is an energy source where the release of CO_2 is balanced by an equal absorption of CO_2 from the atmosphere (e.g. biofuels).

33 The main environmental advantage of nuclear energy is that it does not produce CO_2 and other GHGs.

34 The connections between poverty, vulnerability and disaster risk:

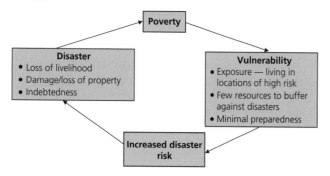

Chapter 2 Going global

1 Foreign direct investment describes inward flows of capital to a country, in the form of plant, machinery, infrastructure etc., mainly as a result of investment by foreign TNCs.

2 TNC = transnational corporation, BRICS = Brazil, Russia, India, China and South Africa, IMF = International Monetary Fund, WB = World Bank, WTO = World Trade Organisation.

3 International migration is the permanent or semi-permanent movement of people from one country to another.

4 Three features of recent internal migration in China are: (1) its unprecedented scale, (2) its rural–urban character, (3) the movement of people from the interior to the coast.

5 The expansion of the EU in 2004 resulted in large-scale net migration to the UK from eastern Europe. The UK, unlike most EU countries, placed no restrictions on immigration from the enlarged EU. The outcome was a huge influx of migrants from Poland and eastern Europe. The attractions of the UK were (a) its labour shortages, and (b) its high-wage economy.

6 The 'North–South' dichotomy is not an accurate description of the geography of global development because (a) many of the world's poorest countries in Sub-Saharan Africa (SSA) and South Asia are in the Northern Hemisphere, and (b) there are several high- and middle-income countries in the Southern Hemisphere, e.g. Australia, New Zealand, Brazil, Argentina.

7 Sub-Saharan Africa is the world's least developed region.

8 LEDC = less economically developed country, MEDC = more economically developed country, LDC = least developed country, NAFTA = North American Free Trade Agreement, NIC = newly industrialising country, OPEC = Organisation of Petroleum Exporting Countries, OECD = Organisation for Economic Cooperation and Development.

9 Free trade allows countries to specialise in producing goods and services for which they have a comparative advantage. In theory this should reduce their cost, thereby stimulating demand and international trade.

10 (1) It creates jobs, (2) introduces new technologies and skills, (3) helps to improve the balance of trade in recipient countries, and (4) lowers costs and gives access to new markets for TNCs.

11 Remittances are monies sent by foreign workers and international immigrants back to their families and relatives in their native countries.

12 (1) International population movements, (2) capital in the form of FDI, (3) international trade in goods and services.

13 Global communications networks facilitate flows of capital, information and services that generate wealth and prosperity. Places isolated from these networks are 'switched off' from the global economy. Isolated, they have fewer opportunities to generate wealth and are therefore more likely to be poor.

14 (a) Offshoring is the relocation of an economic activity by a business from its home country to another country (often where costs are lower). (b) Outsourcing is the contracting out of manufacturing or administrative services by a business to a third party (e.g. sportswear companies contracting-out manufacturing to firms in east Asia). (c) E-commerce refers to trading using modern telecoms (e.g. internet). Examples include Amazon and ebay.

15 Labour costs include wages as well as employers' contributions to pensions, healthcare plans, etc. Unit labour costs relate wages to productivity. Where productivity is high, even if labour costs are rising, unit costs are likely to be low. In China, labour productivity is high owing to manufacturing efficiency and the willingness of employees to work long hours. Also, compared with MEDCs, the responsibilities of employers to provide comfortable working conditions, pension contributions, etc. may be absent.

16 The crude birth rate (CBR) is the average number of children born for every 1,000 of the population. Fertility is the average number of children born to each woman.

17 Civil registration provides demographic information on births, deaths and marriages.

18 Fertility surged in the UK immediately after 1945 because during the war years (1939–1945) couples delayed marrying and/or delayed having children.

19 (1) Advances in medical technology, (2) better sanitation and hygiene, (3) more healthy diets.

20 'Push' factors are the negative factors (economic, social, political, environmental) in a migrant's place of origin; 'pull' factors are the opposite, i.e. the attractive qualities of a migrant's chosen destination. Most migration movements result from a combination of 'push' and 'pull' factors.

21 'Family reunification' means that international immigrants (often young men) can be joined legally in their country of adoption by spouses, children and other close family members.

22 The number of asylum seekers in western Europe rose steeply in the early twenty-first century because wars in Afghanistan, Iraq and Africa created thousands of refugees.

23 International migration of retirees in Europe has been boosted by generous occupational pensions; familiarity with countries such as Spain, Portugal and Italy because of package holidays and mass tourism; house price inflation in their home country (particularly in the UK); and cheap budget airlines that enable retirees to remain in contact with family and friends.

24 The economic benefits of international immigration are that most immigrants are young adults of working age who pay taxes, the relative youthfulness of immigrants helps offset the economic problems of ageing populations, immigrants are often well educated and skilled, immigrants are often prepared to do jobs shunned by the indigenous workforce (e.g. in agriculture).

25 Large-scale immigration can create social problems. Segregation in urban enclaves and policies of multiculturalism can alienate immigrants from the host population. Where immigrant groups are large, they become self-sufficient and integration becomes less of a priority for them. Immigrant groups that lead 'separate lives' may engender prejudice, misunderstanding and hostility in the indigenous population.

26 Million cities have populations of 1 million or more. Mega cities have populations of 10 million or more.

27 (1) Unemployment, (2) poor educational opportunities, (3) poverty.

28 Urbanisation is an increase in the proportion of urban dwellers in a population, suburbanisation is the growth of population and the built area in the contiguous urban zone beyond the inner city, counterurbanisation is an increase in the proportion of rural dwellers in a population, reurbanisation is an increase in the resident population in the central areas of towns and cities.

29 The cycle of urbanisation:

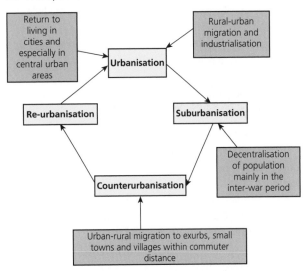

30 Globalisation has often had a bland, homogeneous effect on local cultures, especially in LEDCs, i.e. 'McDonaldisation', 'Americanisation'. Elsewhere, and especially in MEDCs, it has increased cultural diversity, e.g. exotic foods, carnivals, music/films, religious festivals.

31 (1) Increased consumption of unhealthy foods (e.g. foods high in saturated fats, sugary drinks), (2) promotion of tinned milk at the expense of breast feeding, (3) spread of tobacco smoking in LEDCs.

32 Landfill sites are sites where solid waste (both domestic and industrial) is dumped; brownfield sites are vacant sites, most often in urban areas, which have been developed previously but which are now derelict or unused.

33 (1) Recycling of paper, glass, metals, plastics, etc.; (2) green taxes on motor fuel and air passengers, etc.; (3) energy conservation, e.g. domestic insulation, double-glazing; (4) renewable energy, e.g. wind, solar, geothermal.

34 Local sourcing of food (a) provides local employment in food production, processing and transport and (b) reduces 'food miles' and packaging and therefore carbon emissions.

Chapter 3 Extreme weather

1 Tropical cyclones, droughts, tornadoes, blizzards.

2

	Depressions	Anticyclones
Pressure	Low	High
Wind circulation	Anticlockwise	Clockwise
Air masses	Two, separated by fronts	One
Isobars	Close together	Widely spaced
Precipitation	Prolonged	Usually dry

3 Extreme weather conditions brought by depressions include heavy and prolonged spells of precipitation, and strong to gale force winds. Anticyclones often bring long periods of dry weather (droughts). Winter temperatures may be sub-zero for several days, in summer heat wave conditions sometimes occur.

4 (a) Tropical cyclone intensity is measured on the Simpson-Saffir scale, (b) tornado intensity is measured on the Fujita scale.

5 (1) Irrigation, (2) zero tillage, (3) conversion of arable land to grazing, (4) using drought-resistant crops.

Chapter 4 Crowded coasts

1 Cliffs, caves, arches and stacks.

2 The physical factors involved in the formation of coastal dunes are: the prevailing wind direction and its strength, the existence of an extensive backshore area where sand can accumulate, large areas of sand exposed at low tide, and a shallow offshore gradient, maximising the nearshore area exposed at low tide.

3 The rapid growth of population in many coastal areas is due mainly to retirement migration and the development of mass tourism.

4 Storm surges are raised sea levels associated with tropical cyclones, caused by low pressure and hurricane force winds over the sea. Tsunamis are huge waves, caused by seismic movements on the ocean floor. Risks from storm surges might increase in future as tropical cyclones become more frequent and more intense, and more people live in coastal regions exposed to storm surges. The risks from tsunamis might rise as coastal development in earthquake regions increases.

5 Shoreline Management Plans have three strategies: (1) hold-the-line: maintain and in some cases strengthen existing sea defences, (2) no active intervention: allow natural processes to operate without human interference, (3) managed realignment: set back the shoreline and allow the sea to flood areas previously protected by seawalls, embankments, etc.

Chapter 5 Unequal spaces

1 Income, education, housing, employment, crime.

2 Rural housing has been made unaffordable to local people because of (a) influxes of wealthy commuters, (b) rural housing sold for second homes, (c) few new houses built.

3 Rural dwellers without a car and/or where public transport provision is inadequate have difficulty in accessing employment, shopping, medical care and educational services.

4 Economic and social exclusion is a situation where people are prevented from participating fully in the economic, social and political life of society. This may be due to low incomes, unemployment, criminal record, postcode, lack of mobility, etc.

5 Developers, local community members, local councils, housing associations.

Chapter 6 Re-branding places

1 Re-branding is the marketing of a place to give it a new identity. Re-imaging contributes to re-branding by creating booster images of a place that counter negative (or outdated) images and perceptions.

2 Re-branding strategies might be applied to (1) a city that has experienced de-industrialisation and economic decline, (2) a traditional seaside resort planning regeneration, (3) a run-down district of a city undergoing gentrification, e.g. docklands.

3 Among the main stakeholders in re-branding projects are local authorities, government agencies, property developers, land owners and local communities.

4 Flagship developments are large-scale, one-off property projects designed to stimulate further investment and regeneration, e.g. sports stadia, luxury shopping malls. Market-led re-branding involves businesses (e.g. developers) seeking to profit from investment. Examples include gentrification of central London districts such as Notting Hill and Clapham. Top-down re-branding involves partnerships between large public organisations (e.g. councils, development agencies) and private-sector companies. Controlled from the top, it is co-ordinated and well planned, e.g. the regeneration of Salford Quays and Cardiff Bay.

5 (a) Improvements in the quality of the physical environment, (b) visitor numbers, (c) impact on local businesses, (d) visitor spending, (e) footfall in regenerated areas.